Preface

This book presents a course in Linear Programming which will be suitable for a wide range of undergraduate and postgraduate courses. Students of Mathematics, Science and Engineering, as well as students on Management courses which contain a solid component of quantitative methods, will find the text very useful and suited to their needs. Postgraduate students, particularly those on conversion courses in Operational Research or Management Science, will also find the material of value.

The mathematical theory is carefully developed, although there are more rigorous treatments. What the text may lack in rigour is more than compensated for by the main thrust of this book, which is to show, in a vivid manner, the way in which the theoretical ideas are transformed into practical computational procedures capable of implementation on a digital computer. The methods are computer methods and should be treated as such. It is no accident that the main theoretical developments coincided in time with the advent of the computer. Without the latter it would not have been possible to exploit the potential of the theory in the actual solution of real problems.

The text assumes that the reader is familiar with the computer language Pascal. The programs have been written in this language. Of course many main frame computers contain packages which will implement the methods discussed in this book. However, these can be remote and can be treated as a black box and thus used without full understanding. Pascal is available on many microcomputers and their interactive nature will allow the student to become intimately involved with the programs. It is not claimed that these are incapable of improvement. Indeed the authors would be delighted to hear from readers who feel that they can produce better programs. It is through such close involvement that students will gain a real appreciation of the significance of the theory as well as developing their practical computational and programming techniques. The programs given are robust and practical.

The modelling aspects of Linear Programming are catered for through the worked examples within the text, and the exercises to be found at the end of each chapter. The reader is strongly advised to test his skills with these problems. Although some of them are necessarily abbreviated and simplified, they do point to the wide variety of situations in which Linear Programming can find application.

A few remarks about Pascal and the way it is used in this book are appropriate. The programs have been written in Level 0 ISO Pascal. The modular and data structuring facilities of Pascal have been used in implementing the algorithms and these, together with the comments in the code, should make the programs reasonably easy to follow.

Since Pascal is a compiled language, each linear programming problem may require some modification to the program constants and recompilation before the problem can be solved. The parts of the program that may require such modification are indicated by the symbols {**} on the right hand side of the appropriate lines. Alternatively, the reader may prefer to declare these program constants as variables, introducing new constants with sensible upper bounds for any subranges and arrays that exist. Several problems could then be solved by the same compiled program, providing the problem parameters are made part of the input data.

All the programs given in this book have been compiled and run successfully on a mainframe (CDC 180–830 using the University of Minnesota Pascal–6000 V4.0 compiler running under NOS 2.3), a minicomputer (Vax 11/750 using the Berkeley Pascal compiler running under UNIX 4.2BSD) and a microcomputer (an IBM-compatible Ericsson PC using Turbo Pascal running under DOS 2.11). The output shown in the text is that from the CDC machine.

It may be necessary or desirable to make slight alterations to the input/output instructions in the program code dependent on the host machine and whether inter-active features are favoured or data and output files used. For example, on Turbo Pascal using files it will be necessary to assign file names, reset and rewrite files, include file names in read/write instructions and close files. The manner in which the programs have been written should make the points of required alteration easy to identify.

It is a pleasure to thank friends, colleagues and students who have contributed to this work. The students have been willing guinea-pigs for many of the problems. Some of these have been taken from examination papers set at the University of Bradford. We are grateful to the university for permission to use these questions. We are indebted to Mr C. Mack for many beneficial and illuminating discussions on the ideas of his method for the Assignment Problem. Last but not least we must thank Mrs Valerie Hunter for her help in transforming what was at times a messy manuscript into a neat and tidy typescript.

BRIAN BUNDAY AND GERALD GARSIDE
1987

LINEAR PROGRAMMING
IN
PASCAL

Brian D. Bunday,

B.Sc., Ph.D., F.S.S., F.I.M.A.

School of Mathematical Sciences, University of Bradford

Gerald R. Garside,

B.Sc., M.Sc., A.F.I.M.A., M.B.C.S.

School of Computing, University of Bradford

Edward Arnold

© B. D. Bunday 1987

First published 1987 by
Edward Arnold (Publishers) Ltd
41 Bedford Square, London WC1B 3DQ

Edward Arnold
3 East Read Street, Baltimore,
Maryland 21202, USA

Edward Arnold (Australia) Pty Ltd
80 Waverley Road, Caulfield East,
Victoria 3145, Australia

British Library Cataloguing in Publication Data

Bunday, B. D.
   Linear programming in Pascal.
   1. Linear programming—Data processing
   2. PASCAL (Computer program language)
   I. Title   II. Garside, Gerald
   519.7′2′02855133   T57.74

   ISBN 0-7131-3647-2

Text set in 10/12pt Times Compugraphic
by Mathematical Composition Setters Ltd, Salisbury
Printed and bound in Great Britain
by J. W. Arrowsmith, Bristol

Contents

Preface   **(iii)**

1   Fundamental Ideas   **1**
1.1   Introduction   1
1.2   Graphical solution of two-dimensional problems   4
1.3   A standard form for linear programming problems   8
1.4   Some n-dimensional geometry   10
1.5   Fundamental results for linear programming   11

2   The Simplex Method   **19**
2.1   The Simplex Method given an initial basic feasible solution   19
2.2   Implementing the Simplex Method on the computer   26
2.3   Generating an initial basic feasible solution   31
2.4   The full Simplex Method   35
2.5   The problem of degeneracy   43

3   Sensitivity Analysis   **54**
3.1   The inverse of the basis and the simplex multipliers   54
3.2   The effect of changes in the problem   58
3.3   The Dual Simplex method   63

4   The Transportation Problem   **76**
4.1   The nature of the problem and its solution   76
4.2   The 'Stepping Stones' algorithm   81
4.3   Unbalance and degeneracy in the Transportation problem   84
4.4   Implementing the Transportation algorithm on the computer   89

5   The Assignment Problem   **105**
5.1   Introduction   105
5.2   Mack's method of solution   106
5.3   A computer program for Mack's method   110

6   The Revised Simplex Method   **120**
6.1   The Revised Simplex algorithm   120
6.2   Initiating the algorithm   126
6.3   Degeneracy revisited   128
6.4   A computer program for the Revised Simplex Method   131

7	**Duality in Linear Programming**	**148**
7.1	The Primal and Dual problems	148
7.2	Duality theorems	151
7.3	Looking back in the light of duality	157

8	**Linear Programming in Integers**	**164**
8.1	Some problems involving integer programming	164
8.2	Gomory's method for all integer linear programming	166
8.3	A computer program for Gomory's method	169
8.4	Dantzig's method	176

Suggestions for Further Reading	**180**
Solutions	**181**
Index	**187**

1

Fundamental Ideas

1.1 Introduction

Linear programming has proved itself to be an extremely powerful technique in the solution of certain problems which arise in the field of Operational Research. The word 'programming' means planning in this context, and gives a clue as to the type of application. The ideas were first developed during World War II in connection with finding optimal strategies for conducting the war effort. Since that time they have found wide application in industry, commerce and Government Service, the latter at both local and national level. The methods are of value in the formulation and solution of many (though not all), problems concerned with the efficient use of limited resources.

Some of the ideas can be illustrated from consideration of the following simplified version of a real production scheduling problem.

Example 1

A firm produces self-assembly bookshelf kits in two models, A and B. Production of the kits is limited by the availability of raw material (high quality board) and machine processing time. Each unit of A requires 3 m^2 of board and each unit of B requires 4 m^2 of board. The firm can obtain up to 1700 m^2 of board each week from its suppliers. Each unit of A needs 12 minutes of machine time and each unit of B needs 30 minutes of machine time. Each week a total of 160 machine hours is available. If the profit on each A unit is $2, and on each B unit is $4, how many units of each model should the firm plan to produce each week?

In order to formulate this problem in mathematical form let the weekly production of A be x_1 units, and of B, x_2 units. The problem is then to find the *best* values for x_1 and x_2. A fairly obvious way to interpret best for this problem is *so as to maximise profit each week*. The weekly profit can be expressed as

$$P = 2x_1 + 4x_2. \tag{1.1}$$

The firm will achieve its objective by maximising the **objective function** $P = 2x_1 + 4x_2$.

Classical optimisation says that an optimum of a function will occur *either* where the derivatives are zero *or* on the boundary of the domain space. To consider the derivatives only is inadequate.

$$\frac{\partial P}{\partial x_1} = 2 \quad \text{and} \quad \frac{\partial P}{\partial x_2} = 4$$

and it is not possible to make these derivatives zero by choice of x_1 and x_2. Indeed the way to increase P is to go on increasing x_1 and x_2. But, and this is the essence of the problem, x_1 and x_2 cannot be increased without limit. Their possible values are restricted by physical considerations and by the **constraints** on raw material and machine time.

Because x_1 and x_2 represent the number of units of each model produced each week, it is clear that they cannot be negative:

$$\text{i.e. } x_1 \geqslant 0, x_2 \geqslant 0. \tag{1.2}$$

The constraints on availability of board and machine time can be put in the form:

$$\text{Board:} \qquad 3x_1 + 4x_2 \leqslant 1700$$
$$\text{Machine hours:} \quad \tfrac{1}{5}x_1 + \tfrac{1}{2}x_2 \leqslant 160$$

$$\text{i.e. } \left.\begin{array}{r} 3x_1 + 4x_2 \leqslant 1700 \\ 2x_1 + 5x_2 \leqslant 1600 \end{array}\right\}. \tag{1.3}$$

Thus the problem is to find values of x_1 and x_2 which satisfy the **non-negativity** conditions (1.2) and the inequality constraints (1.3) so as to maximise $P = 2x_1 + 4x_2$.

This is a typical linear programming problem. The objective function which is to be maximised is a **linear function** of the variables. The constraints on these variables are also linear. Indeed for this particular two-dimensional problem they can be represented graphically by the lines shown in Fig. 1.1. The non-negativity conditions restrict the variables to the positive quadrant. The constraints are represented by the lines:

$$3x_1 + 4x_2 = 1700$$
$$2x_1 + 5x_2 = 1600.$$

The arrow on each constraint in Fig. 1.1 indicates the side of the line on which the constraint is satisfied. The directions on each arrow can easily be determined by considering whether the origin $(0, 0)$ satisfies the constraint. The shaded area OABC which contains all points (x_1, x_2) satisfying equations (1.2) and (1.3) is called the **feasible region**. Points within and on the boundary of this region represent **feasible solutions** of the constraints. There are plenty of feasible solutions. The problem is to find the one (or might there be more than one?) which maximises P.

The (dashed) lines (a) $2x_1 + 4x_2 = 0$, (b) $2x_1 + 4x_2 = 800$, are shown in Fig. 1.1. These lines are parallel and represent two **contour** lines of the function P with values 0 and 800 respectively. It is clear that the value of P increases as the contour lines move further away from the origin in the positive quadrant. Indeed the vector with

components $\begin{pmatrix} \dfrac{\partial P}{\partial x_1} \\[2mm] \dfrac{\partial P}{\partial x_2} \end{pmatrix}$ i.e. the vector with components $\begin{pmatrix} 2 \\ 4 \end{pmatrix}$ points in the direction of

increasing P and this direction is perpendicular to these parallel lines, away from 0, as shown.

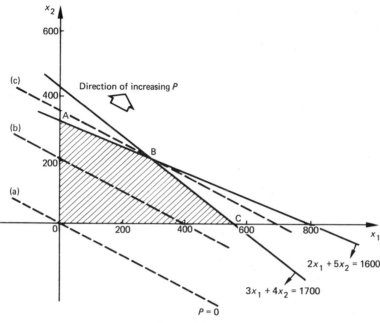

Figure 1.1

The contour line with the highest value of P which has at least one point in common with the feasible region is the line (c), which passes through the vertex B, and on which P has the value 1400. The point B, $x_1 = 300$, $x_2 = 200$ corresponds to the optimal solution for the problem. These values can be obtained as the solution of the equations

$$3x_1 + 4x_2 = 1700$$
$$2x_1 + 5x_2 = 1600.$$

Of course the maximum profit is then $2 \times 300 + 4 \times 200 = 1400$. In the optimal solution both constraints are satisfied as equalities, and this can be interpreted as meaning that all the available raw material and machine time is utilised.

It is clear that this problem could be extended. There could be three or more models and a corresponding number of non-negative variables. There could be additional constraints representing market capacity, limitations on packaging facilities, etc. The problem would still be one of maximising a *linear* function of several *non-negative* variables which are subject to *linear* inequality constraints.

The general linear programming problem is that of maximising (or it could be minimising) a linear function

$$z = c_1x_1 + c_2x_2 + \cdots + c_nx_n \tag{1.4}$$

of n real variables $x_1, x_2, \cdots, x_n$ satisfying the non-negativity conditions

$$x_1 \geqslant 0, x_2 \geqslant 0, ..., x_n \geqslant 0 \tag{1.5}$$

and m linear constraints

$$\left.\begin{array}{l} a_{11}x_1 + a_{12}x_2 + \cdots + a_{1n}x_n \leqslant = \geqslant b_1 \\ a_{21}x_1 + a_{22}x_2 + \cdots + a_{2n}x_n \leqslant = \geqslant b_2 \\ \cdots\cdots\cdots\cdots\cdots\cdots\cdots\cdots\cdots\cdots \\ a_{m1}x_1 + a_{m2}x_2 + \cdots + a_{mn}x_n \leqslant = \geqslant b_m \end{array}\right\}. \tag{1.6}$$

The constraints can be a mixture of the '$\leqslant$', '$=$' or '$\geqslant$' variety. The aim will be to maximise the objective or perhaps to minimise the objective (if it represents a cost for example). The values of the b_i, c_j, a_{ij} are assumed to be known constants. They will often have a physical interpretation in terms of a practical problem as in Example 1.

In matrix notation the problem can be written

Maximise (or minimise)

$$z = c^{\mathrm{T}}x_0 \tag{1.7}$$

where

$$x_0 \geqslant 0 \tag{1.8}$$

and

$$A_0 x_0 \leqslant = \geqslant b \tag{1.9}$$

where $x_0 = \begin{pmatrix} x_1 \\ x_2 \\ \vdots \\ x_n \end{pmatrix}$ is an $n \times 1$ column vector,

$c^{\mathrm{T}} = (c_1, c_2, \ldots, c_n)$ is a $1 \times n$ row vector,

$b = \begin{pmatrix} b_1 \\ b_2 \\ \vdots \\ b_m \end{pmatrix}$ is an $m \times 1$ column vector, which can be assumed non-negative,

and $A_0 = (a_{ij})$ is an $m \times n$ matrix.

The suffix 0 or x_0 and A_0 is to be taken to mean 'original'. The point will be taken up and made clear in Section 1.3.

1.2   Graphical Solution of Two-Dimensional Problems

The two-dimensional example of the previous section served to show how a linear programming problem could arise from a practical situation and also showed a graphical method of solution. Through a study of a few other such examples which are simple enough for us to see 'what is going on', it will be possible to bring out certain general features of linear programming problems whose exploitation can lead to a systematic solution procedure.

Example 1
Minimise
$$z = -3x_1 - 4x_2$$

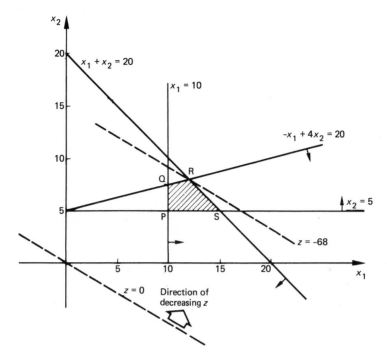

Figure 1.2

where $x_1, x_2 \geqslant 0$ and

$$x_1 + x_2 \leqslant 20$$
$$-x_1 + 4x_2 \leqslant 20$$
$$x_1 \geqslant 10$$
$$x_2 \geqslant 5.$$

PQRS is the feasible region in Fig. 1.2. The last two constraints subsume the non-negativity conditions. z *decreases* in the direction

$$-\begin{pmatrix} \dfrac{\partial z}{\partial x_1} \\[2mm] \dfrac{\partial z}{\partial x_2} \end{pmatrix}, \text{ i.e. } \begin{pmatrix} 3 \\ 4 \end{pmatrix}.$$

The minimum value of z is -68 and arises at R(12, 8). Note that as in the example of the last section the minimum occurs at a vertex of the feasible region. The optimal solution is $x_1 = 12$, $x_2 = 8$ with the minimum of z at -68.

Sometimes there is more than one optimal solution.

Example 2

Minimise $\qquad\qquad\qquad\qquad z = -6x_1 - 2x_2$

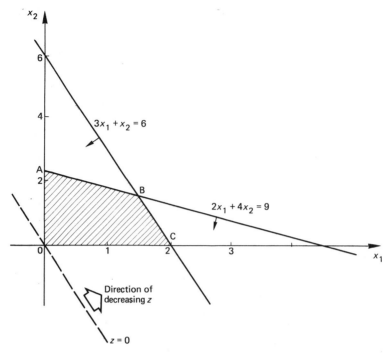

Figure 1.3

subject to $x_1, x_2 \geqslant 0$

$$2x_1 + 4x_2 \leqslant 9$$
$$3x_1 +\ x_2 \leqslant 6$$

OABC in Fig. 1.3 shows the feasible region.

$\partial z/\partial x_1 = -6$, $\partial z/\partial x_2 = -2$ and so the vector $\begin{pmatrix} 6 \\ 2 \end{pmatrix}$ points in the direction of *decreasing z*. Any point on BC represents an optimal solution. In particular the vertices $B(1\frac{1}{2}, 1\frac{1}{2})$ and $C(2, 0)$ represent optimal solutions corresponding to the (one) minimum value of $z = -12$.

Any point on BC can be represented as

$$\theta(1\tfrac{1}{2}, 1\tfrac{1}{2}) + (1 - \theta)(2, 0) = (2 - \tfrac{1}{2}\theta, 1\tfrac{1}{2}\theta)$$

for $0 \leqslant \theta \leqslant 1$.

For each point the value of z is $-6(2 - \tfrac{1}{2}\theta) - 2(1\tfrac{1}{2}\theta) = -12$. There is only *one* minimum value of z.

Sometimes the solution is unbounded.

Example 3
 Maximise $z = x_1 + x_2$

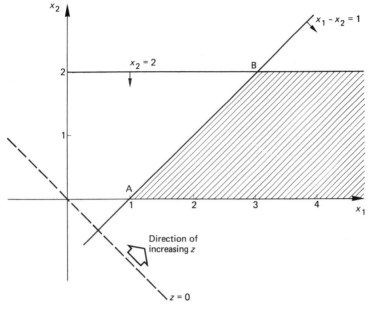

Figure 1.4

subject to $x_1 \geqslant 0$, $x_2 \geqslant 0$

$$x_1 - x_2 \geqslant 1.$$

$$x_2 \leqslant 2.$$

The feasible region shown in Fig. 1.4 is unbounded in the direction in which z increases. There is no finite point in the feasible region at which z attains a maximum. The solution is unbounded and so is the maximum value z. It is possible in some problems for an unbounded solution to occur with a finite maximum for the objective. This would be the case, for example, had the problem been to maximise $z' = x_2$ subject to the constraints.

Of course, had the problem been to *minimise* $z = x_1 + x_2$ subject to the above constraints, there is one finite minimum of z (min) $= 1$ at **A** (again a vertex of the feasible region) where $x_1 = 1$, $x_2 = 0$.

Sometimes there is no solution at all because a feasible region does not exist.

Example 4

Minimise $$z = 2x_1 + 3x_2$$

subject to $x_1, x_2 \geqslant 0$

$$x_1 + \ x_2 \geqslant 10$$
$$3x_1 + 5x_2 \leqslant 15.$$

The constraints are contradictory and have no feasible solution (see Fig. 1.5).

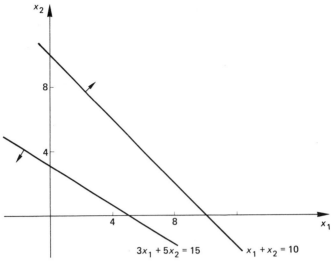

Figure 1.5

One or two fairly general features of linear programming (L.P.) problems can be deduced from the examples already considered. The first is that the feasible region is always a convex polygon. Even in the case where it was unbounded it was convex on its closed aspect. The second is that the optimum solution always occurs at a vertex of the feasible region. In Example 2 where there were several optimal solutions both the vertices B and C corresponded to optimal solutions.

We shall see that we can generalise these results. First we show that we can put all L.P. problems in a standard form.

1.3 A Standard Form for Linear Programming Problems

It may appear that L.P. problems can take on a variety of forms with the constraints a mixture of '$\geqslant$', '$=$' or '$\leqslant$' type. They can all be put into a standard form in which the objective function is to be **minimised** and all constraints take the form of **equations** in **non-negative variables**.

Problems which do not initially conform to this standard form can be brought to this form quite simply.

(a) Maximising the objective function $z = c_1 x_1 + \cdots + c_n x_n$ is equivalent to minimising the objective function

$$z' = - c_1 x_1 - c_2 x_2 - \cdots - c_n x_n.$$

(b) Inequality constraints.

The constraint $3x_1 + 2x_2 - x_3 \leqslant 6$ can be put in equation form as

$$3x_1 + 2x_2 - x_3 + x_4 = 6$$

where the **slack variable** x_4 is non-negative.

The constraint $x_1 - x_2 + 3x_3 \geqslant 10$ can be put in equation form as

$$x_1 - x_2 + 3x_3 - x_5 = 10$$

where the **slack variable** x_5 is non-negative.
(c) Non-negative variables.
   If a particular variable x_k can take on any value then we can write $x_k = x_k' - x_k''$ where $x_k' \geqslant 0$ and $x_k'' \geqslant 0$.

   Thus bringing a problem into the standard form might involve the introduction of additional variables, (which are still non-negative) into the problem.
   Thus following on from equations (1.7), (1.8), (1.9) our most general L.P. problem can be put in the form

minimise $$z = c^{\mathrm{T}}x \tag{1.10}$$

where $$x \geqslant 0 \tag{1.11}$$

and $$Ax = b, \quad \text{with} \quad b > 0. \tag{1.12}$$

If this problem did indeed arise from that given earlier then x will contain the slack variables as well as the *original* variables and A will contain the coefficients of the slack variables as well as the original coefficients.
   Thus Example 1 of Section 1.2 can be put in the form

minimise $$z = -3x_1 - 4x_2$$

subject to the constraints

$$
\begin{aligned}
x_1 \quad\quad - x_3 \quad\quad\quad\quad &= 10 \\
x_2 \quad\quad - x_4 \quad\quad\quad &= 5 \\
x_1 + \ x_2 \quad\quad\quad + x_5 \quad\quad &= 20 \\
-x_1 + 4x_2 \quad\quad\quad\quad + x_6 &= 20
\end{aligned}
$$

and $x_i \geqslant 0, i = 1, 2, ..., 6$.
   Example 1 of Section 1.1 can be put in the form

minimise $$z = -2x_1 - 4x_2$$

subject to the constraints

$$
\begin{aligned}
3x_1 + 4x_2 + x_3 \quad\quad &= 1700 \\
2x_1 + 5x_2 \quad\quad + x_4 &= 1600
\end{aligned}
$$

and $x_i \geqslant 0, i = 1, ..., 4$.
   In matrix form the constraints can be written

$$
\begin{pmatrix} 3 & 4 & 1 & 0 \\ 2 & 5 & 0 & 1 \end{pmatrix}
\begin{pmatrix} x_1 \\ x_2 \\ x_3 \\ x_4 \end{pmatrix}
= \begin{pmatrix} 1700 \\ 1600 \end{pmatrix}
$$

   They consist of 2 equations in 4 unknowns. Any non-negative solution of these constraints is a **feasible solution**.

Of course with 2 equations in 4 unknowns we can hope to get a solution (though not necessarily a feasible solution) by giving two of the variables arbitrary values and solving for the remaining two. Of particular interest are solutions of this type which arise by setting two of the variables to zero. Such a solution, if unique, is referred to as a **basic** solution. If it is also feasible it is a **basic feasible solution** (**b.f.s.**). For the general linear programming problem with say m linear equation constraints in n variables ($m < n$) a **basic solution** of the constraints is obtained by setting ($n - m$) of the variables to zero and solving the m equations which result for the remaining m variables, provided these equations have a **unique** solution. The variables put equal to zero are called the **non-basic variables (n.b.v.)**. The others are the **basic** variables and form a **basis**.

For the problem just considered we can select the two non-basic variables in $\binom{4}{2} = 6$ ways. The basic solutions are easily seen to be given by

	x_1	x_2	x_3	x_4	
1	0	0	1700	1600	O
2	0	425	0	− 525	
3	0	320	420	0	A
4	$566\frac{2}{3}$	0	0	$466\frac{2}{3}$	C
5	800	0	− 700	0	
6	300	200	0	0	B

corresponding to non-basic variables (x_1, x_2), (x_1, x_3), (x_1, x_4), (x_2, x_3), (x_2, x_4), (x_3, x_4). Of these 6 basic solutions, only 4 are also feasible, and it will be seen that these solutions correspond to the vertices of the feasible region in Fig. 1.1 with correspondence as indicated.

In three dimensions the linear constraints take the form of planes (instead of lines). The feasible region, instead of being a convex polygon, is a convex polyhedron. An optimal solution to the problem will correspond to a vertex of this polyhedron since the contours of the objective function will be planes instead of lines, and the plane corresponding to the least value will generally have just one point in common with the feasible region, and this will be a vertex of the convex polyhedron, and will correspond to an optimal solution of the problem.

We shall see that this particular type of result is quite general. The basic feasible solutions of a system of m equations in n unknowns correspond to the vertices of the feasible region. Further an optimal solution, if it exists, corresponds to a basic feasible solution, and hence a vertex of the feasible region.

1.4 Some *n*-Dimensional Geometry

Before establishing the results just mentioned it is necessary to generalise some geometric concepts from two dimensions to n dimensions. The two-dimensional graphical solution method used in Section 1.2 is quite general. However, in n dimensions our intuition and ability to visualise the situation is not so clear. We need algebraic methods to do the geometry.

We first define some terms which allow the concept of a convex set to be understood.

We use

$$x = \begin{pmatrix} x_1 \\ x_2 \\ \vdots \\ x_n \end{pmatrix} \text{ to denote a point in } n\text{-dimensional space.}$$

The **line segment** PQ, where P and Q, represented by vectors p and q, are two points, consists of all points of the form

$$\theta p + (1 - \theta)q; \quad 0 \leqslant \theta \leqslant 1.$$

A set of points S is a **convex set,** if, given that P and Q belong to the set, then so do all points of the line segment PQ. An **extreme point** (**vertex** or **corner**) of a convex set is a point of the set which does not lie on a line segment of any two other points of the set.

The **convex hull** of the points $P_1, P_2, ..., P_k$ with vectors $p_1, p_2, ..., p_k$ is the set of points with vectors given by

$$\theta_1 p_1 + \theta_2 p_2 + \cdots + \theta_k p_k$$

where $\theta_i \geqslant 0$ ($i = 1, 2, ..., k$) and $\sum_{i=1}^{k} \theta_i = 1$.

Figure 1.6(a) represents a convex set. The set in Fig. 1.6(b) is not a convex set. Some points of VW are not in the set. $P_1 P_2 P_3 P_4 P_5$ are the vertices of the first set. The convex hull of two points $P_1 P_2$ is the line segment $P_1 P_2$. The convex hull of three points is the triangle $P_1 P_2 P_3$, of 4 points the tetrahedron $P_1 P_2 P_3 P_4$ and of five points the hyperpolyhedron with the five points as vertices.

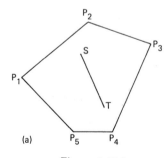

Figure 1.6(a)

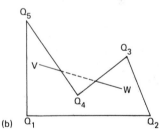

Figure 1.6(b)

1.5 Fundamental Results for Linear Programming

In standard form the general linear programming problem takes the form

minimise
$$z = c_1 x_1 + c_2 x_2 + \cdots + c_n x_n$$

subject to

$$a_{11}x_1 + a_{12}x_2 + \cdots + a_{1n}x_n = b_1$$
$$a_{21}x_1 + a_{22}x_2 + \cdots + a_{2n}x_n = b_2$$
$$\cdots\cdots\cdots\cdots\cdots\cdots\cdots\cdots\cdots\cdots$$
$$a_{m1}x_1 + a_{m2}x_2 + \cdots + a_{mn}x_n = b_m$$

and $x_1, x_2, ..., x_n \geqslant 0$.

The constraints take the form $Ax = b, x \geqslant 0$ where A has rank $m(<n)$.

I   *If the Constraints have a Feasible Solution they have a Basic Feasible Solution*

We can prove this by constructing the b.f.s. Suppose that in the feasible solution $n - r$ variables are zero, the rest positive. Then without loss of generality we have

$$x_j = 0, \quad j = r + 1, ..., n$$

$$\sum_{j=1}^{r} x_j a_j = b, \quad \text{where} \quad x_j > 0 \quad \text{for} \quad j = 1, 2, ..., r, \tag{1.13}$$

and the a_j are the columns of A.

If these are linearly independent then $r \leqslant m$, the rank of A, and the solution is a basic feasible solution ($m - r$ of the *basic* variables are zero).

If $a_1, a_2, ..., a_r$ are not linearly independent then

$$\sum_{j=1}^{r} \alpha_j a_j = 0 \text{ for some } \alpha_1, ..., \alpha_r \quad \text{not all zero or negative.} \tag{1.14}$$

$$\text{(if so multiply by } -1)$$

Thus if $\alpha_k > 0$,

$$a_k = -\sum_{\substack{j \neq k \\ j=1}}^{r} \frac{\alpha_j a_j}{\alpha_k} \tag{1.15}$$

so that

$$\sum_{j=1}^{r} \left(x_j - \frac{\alpha_j x_k}{\alpha_k} \right) a_j = b.$$

Now if we choose k so that

$$\frac{x_k}{\alpha_k} = \min_{j=1...r} \left(\frac{x_j}{\alpha_j}; \ \alpha_j > 0 \right)$$

then the values

$$\left. \begin{array}{l} X_j = x_j - \alpha_j \left(\dfrac{x_k}{\alpha_k} \right), \ j = 1, ..., r(j \neq k), \\[2mm] X_j = 0, \qquad\qquad\quad j = k, r+1, ..., n, \end{array} \right\} \tag{1.16}$$

are non-negative and so form a feasible solution with at most $r - 1$ strictly positive values. Thus we have decreased the number of positive variables by one. Continuing in this way we shall eventually reach a state when $r \leqslant m$, in which case we have a basic feasible solution.

II *The Feasible Region is a Convex Set*

If x and y are in the feasible region so that $Ax = b$ and $Ay = b$ and $x, y \geqslant 0$ then if

$$w = \theta x + (1 - \theta)y \quad \text{where} \quad 0 \leqslant \theta \leqslant 1,$$

then $w \geqslant 0$.

Also

$$Aw = A(\theta x + (1 - \theta)y) = \theta Ax + (1 - \theta)Ay$$
$$= \theta b + (1 - \theta)b$$
$$= b.$$

Thus w is in the feasible region which is hence a convex set.

III *Basic Feasible Solutions Correspond to Vertices of the Convex Set*

Suppose $x^0 = \begin{pmatrix} x_1 \\ x_2 \\ \vdots \\ x_m \\ 0 \\ 0 \\ 0 \end{pmatrix}$ is a basic feasible solution. Thus x^0 is the **unique** solution

of $Ax = b; x \geqslant 0$ with zeros in the last $n - m$ positions as shown.

If x^0 is *not* a vertex, then two *other* points u and v can be found such that $x^0 = \theta u + (1 - \theta)v$ for some θ, where $0 < \theta < 1$, and where $Au = b, Av = b, u, v \geqslant 0$.

Thus for the last $n - m$ components we have

$$\theta u_{m+1} + (1 - \theta)v_{m+1} = 0$$
$$\dots \dots \dots \dots \dots \dots \dots$$
$$\theta u_n + (1 - \theta)v_n = 0.$$

But since, $\theta, 1 - \theta, u_j$ and $v_j \geqslant 0 (j = 1, 2, ..., n)$ the only way in which

$$\theta u_j + (1 - \theta)v_j = 0 \quad (j = m + 1, ..., n)$$

is if $u_j = v_j = 0$ for $j = m + 1, ..., n$.

Thus u and v are also basic feasible solutions with the same zeros as x^0. Thus since x^0 is unique we deduce that $x^0 = u = v$ which is a contradiction, so that each basic feasible solution is a vertex.

We can also prove that the converse of this is true, i.e. the vertices correspond to the basic feasible solutions.

Let x^0 be a vertex of the feasible region. Suppose x^0 has r strictly positive components. We shall show that r is at most m so that x^0 is a b.f.s. Suppose $x_1^0, x_2^0, ..., x_r^0 (r > m)$ are positive. Let $a_1, a_2, ..., a_r$ be the columns of A corresponding to these and suppose that they are linearly dependent.

Then as in I there will be α_j, not all zero, such that

$$\sum_{j=1}^{r} \alpha_j a_j = 0.$$

Now it is easy to see that provided

$$0 < \rho < \min_{j} \frac{x_j^0}{|\alpha_j|} \quad \text{for} \quad \alpha_j \neq 0,$$

then the vectors

$$x_1 = x^0 + \rho\alpha, \quad x_2 = x^0 - \rho\alpha$$

where

$$\alpha = \begin{pmatrix} \alpha_1 \\ \alpha_2 \\ \alpha_r \\ 0 \\ 0 \end{pmatrix}$$

are such that $x_1, x_2 \geqslant 0$.

Further since $A\alpha = 0$

$$Ax_1 = A(x^0 + \rho\alpha) = Ax^0 + \rho A\alpha = b$$

Similarly $Ax_2 = b$.

Thus x_1 and x_2 are feasible solutions and $x^0 = \frac{1}{2}(x_1 + x_2)$ so that x^0 is not a vertex. This then leads to a contradiction and we conclude that r is at most m.

If there are m constraints in n variables, there are at most $\binom{n}{m}$ basic feasible solutions (and vertices) and the convex hull of these points forms the feasible region.

IV *If the Objective Function has a Finite Minimum, then at Least One Optimal Solution is a Basic Feasible Solution*

Let the basic feasible solutions correspond to the points $P_1, P_2, ..., P_k$ with vectors $p_1, p_2, ..., p_k$, and suppose the objective function takes the values $z_1, z_2, ..., z_k$ at these points.

If $z = c_1 x_1 + c_2 x_2 + ... + c_n x_n = c^T x$

$$z_i = c^T p_i, \quad \text{for} \quad i = 1, 2, ..., k.$$

For any other point in the feasible region

$$x = \theta_1 p_1 + \theta_2 p_2 + ... + \theta_k p_k \quad \text{where} \quad \theta_i \geqslant 0, \quad \Sigma\theta_i = 1.$$

The value of z at this point is

$$z = c^T x = \theta_1 c^T p_1 + \theta_2 c^T p_2 + ... + \theta_k c^T p_k$$
$$= \theta_1 z_1 + \theta_2 z_2 + ... + \theta_k z_k.$$

Thus the problem of finding x in the convex hull of $P_1, P_2, ..., P_k$ which minimises z, reduces to the problem of finding $\theta_i \geqslant 0$ such that

$$\sum_{i=1}^{k} \theta_i = 1,$$

which minimises

$$\sum_{i=1}^{k} \theta_i z_i.$$

Now of the values $z_1, z_2, ..., z_k$ there will be one (or more) that is the minimum. Suppose this is z_j so that $z_j \leqslant z_i$ for $i = 1, 2, ..., k$.

Thus $\Sigma \theta_i z_i$ which is the weighted average of $z_1 ... z_k$ with weights $\theta_1, \theta_2, ..., \theta_k$ will be a minimum when $\theta_j = 1$ and $\theta_i = 0 (i \neq j)$. Thus the minimum of z will arise at the vertex P_j.

These results mean that in searching for the optimal solution in the feasible region it is only necessary to consider the basic feasible solutions. The Simplex Method of the next chapter is a systematic procedure for doing this.

Exercises 1

1 A firm manufactures two products A and B, the market for each being virtually unlimited. Each product is processed on each of the machines I, II and III. The processing times in hours per item of A or B on each machine are given in the table.

	I	II	III
A	0.5	0.4	0.2
B	0.25	0.3	0.4

The available production time of the machines I, II and III is 40 hours, 36 hours and 36 hours respectively each week. The profit per item of A and B is $5 and $3 respectively.

The firm wishes to determine the weekly production of items of A and B which will maximise its profit. Formulate this problem as a linear programming problem and solve it.

2 Maximise
$$w = x_1 + 2x_2$$
subject to $x_1 \geqslant 0, x_2 \geqslant 0$

$$-x_1 + 3x_2 \leqslant 10$$
$$x_1 + x_2 \leqslant 6$$
$$x_1 - x_2 \leqslant 3$$
$$x_1 + 4x_2 \geqslant 4.$$

3 Express the problem above in standard form. Show that there are 15 basic solutions of which 5 are also feasible. Identify these with the vertices of the feasible region.

4 Minimise
$$z = -2x_1 - x_2$$
subject to $x_1 \geqslant 0, x_2 \geqslant 0$

$$x_1 + 2x_2 \leqslant 11$$
$$x_1 + x_2 \leqslant 6$$
$$x_1 - x_2 \leqslant 2$$
$$2x_1 - 4x_2 \leqslant 3.$$

5   Minimise $$z = -3x_1 - x_2$$

subject to $x_1 \geqslant 0$, $x_2 \geqslant 0$

$$
\begin{aligned}
x_1 + x_2 &\geqslant 1 \\
x_1 - x_2 &\leqslant 1 \\
-2x_1 + x_2 &\leqslant 3 \\
\alpha x_1 + \beta x_2 &\leqslant 6
\end{aligned}
$$

in each of the cases

(a) $\alpha = \beta = 1$ (b) $\alpha = 2, \beta = \frac{2}{3}$ (c) $\alpha = 6, \beta = -6$.

6   Minimise $$z = -x_1 - 5x_2$$

subject to $x_1, x_2 \geqslant 0$

$$
\begin{aligned}
x_1 + x_2 &\geqslant 6 \\
3x_1 + 4x_2 &\leqslant 12.
\end{aligned}
$$

7   Maximise $3x_1 + 6x_2 + 2x_3$ where $x_1, x_2, x_3 \geqslant 0$ and

$$
\begin{aligned}
3x_1 + 4x_2 + x_3 &\leqslant 2 \\
x_1 + 3x_2 + 2x_3 &\leqslant 1.
\end{aligned}
$$

For this three-dimensional problem show that the feasible region is a convex polyhedron and that the optimal solution occurs at a vertex.

8   A firm requires coal with a phosphorus content no more than $0 \cdot 03\%$ and no more than $3 \cdot 25\%$ ash impurity. Three grades of coal A, B, C are available at the prices shown.

Grade	% Phosphorus	% Ash	Cost($/tonne)
A	$0 \cdot 06$	$2 \cdot 0$	30
B	$0 \cdot 04$	$4 \cdot 0$	30
C	$0 \cdot 02$	$3 \cdot 0$	45

How should these be blended to meet the impurity restrictions at minimum cost? [Hint. If we let one tonne of the blend contain x_1, x_2, x_3 tonnes of A, B, C respectively, $x_1 + x_2 + x_3 = 1$. Thus this three-dimensional problem can easily be reduced to two-dimensional form by eliminating x_3.]

9   Floor cleaning detergents for domestic usage are formulated with three characteristics in mind, (a) cleaning capacity, (b) disinfectant power, (c) skin irritation level. Each of these characteristics is measured on a linear scale 0–100.

The market product must have a cleaning capacity of at least 60 units and disinfectant power of at least 60 units as measured on these scales. At the same time it is desired to minimise the skin irritation level. This final product is to be a blend of three basic detergents whose characteristics are detailed below.

Basic detergent	Cleaning capacity	Disinfectant power	Irritation level
A	90	30	70
B	65	85	50
C	45	70	10

Formulate the problem of finding the optimum blend as a linear programming problem. Show that this problem can be represented graphically in *two* dimensions and hence obtain the solution.

10   A firm manufactures two products A and B which are sold at 8 cents and 15 cents per unit respectively, the market for both products being virtually unlimited. A is processed on machine I, B is processed on machine II. Then both are packaged at the packing plant (see sketch).

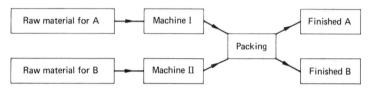

Raw material costs 6 cents per kg and is processed at 5000 kg per hour by machine I with 10% wastage; it is processed at 4000 kg per hour by machine II and 20% is wasted. Machine I is available 6 hours a day with a running cost of $288 per hour; machine II is available 5 hours a day with a running cost of $336 per hour. Finished units of A weigh $\frac{1}{4}$ kg and units of B weigh $\frac{1}{3}$ kg. The packing plant is available 10 hours a day at a cost of $360 per hour. Units of A can be packed at 12 000 per hour whilst units of B are packed at the slower rate of 8000 units per hour.

The company wishes to determine those values for x_1 and x_2, the input (in thousands of kg) of raw material used for the products A and B respectively, so as to maximise daily profit. Formulate this problem as a linear programming problem and calculate the optimal solution graphically.

11   For the two-dimensional examples of Section 1.2, put each problem into standard form. Find all the basic solutions and show that the basic feasible solutions can be identified with the vertices of the feasible region.

12   In a certain area the need arises for additional transport at two points A and B. Five extra buses are required at A and seven extra buses at B. It is known that 3, 4 and 5 buses can be allocated from three garages G_1, G_2 and G_3 respectively.

How should these buses be distributed between A and B in order to minimise the total distance travelled?

The distances from garages to A and B are given in the following table:

Distance	A	B
G_1	3	4
G_2	1	3
G_3	4	2

13   A company produces bathroom cabinets in 2 sizes A and B. Their sales staff tell them that the market can absorb up to 550 cabinets per week. Each cabinet of type A needs 2 m^2 of material and each cabinet of type B needs 3 m^2 of material. The company can obtain up to 1200 m^2 of material weekly. Each A cabinet needs 12 minutes of machine time and each B cabinet 30 minutes of machine time. Each week 160 hours of machine time is available. If the profit on each A cabinet is $3, and on each B cabinet $4, how many of each should be made each week?

14   A car factory produces two models, the Caprice and the (cheaper) Fiasco. The factory has 1000 unskilled workers and 800 skilled workers, each of whom is paid for a 40-hour week. A Caprice requires 30 hours of unskilled labour and 50 hours of skilled labour. A Fiasco requires 40 hours of unskilled and 20 hours of skilled labour. Each Fiasco requires an outlay of $500 for raw materials and parts, whilst each Caprice requires an outlay of $1500; the total outlay must not exceed $900 000 per week. The delivery workers work a five-day week and can only remove 210 cars per day from the factory.

The firm makes a profit of $1000 on each Caprice and $500 on each Fiasco. What output of each model would you recommend?

What possibilities would you suggest for improving the profitability of the firm?

15   A firm has factories in Leeds and Cardiff which supply goods to depots in Manchester, Birmingham and London. The distances between these towns (to the nearest ten miles) are shown in the table below.

	Manchester	Birmingham	London
Leeds	40	110	190
Cardiff	170	100	150

The Leeds factory has an output of 800 tonnes per year and the Cardiff factory an output of 500 tonnes. The Manchester depot requires 400 tonnes, the Birmingham depot 600 tonnes and the London depot 300 tonnes. How should the goods be transported to minimise the transport costs?

In a particular year, there are major roadworks between London and Cardiff which double the cost per tonne-mile on this route. How would you revise your schedule?

2
The Simplex Method

2.1 The Simplex Method given an Initial Basic Feasible Solution

The graphical method of Section 1.2, although useful for 2-dimensional problems, becomes difficult (to draw) for 3-dimensional problems and impossible in other cases. However, the optimum solution is given by a basic feasible solution in all problems. The Simplex Method, developed by George Dantzig, is a computational procedure which exploits this result, but in algebraic form.

It can be applied immediately to the general linear programming problem in standard form:

$$\text{minimise} \qquad c_1 x_1 + c_2 x_2 + \cdots + c_n x_n = \boldsymbol{c}^T \boldsymbol{x} = z \qquad (2.1)$$

$$\text{subject to} \qquad x_1, x_2, \ldots, x_n \geqslant 0; \qquad \text{i.e.} \quad \boldsymbol{x} \geqslant \boldsymbol{0}, \qquad (2.2)$$

$$\text{and} \qquad \left. \begin{array}{l} a_{11} x_1 + a_{12} x_2 + \cdots + a_{1n} x_n = b_1 \\ a_{21} x_1 + a_{22} x_2 + \cdots + a_{2n} x_n = b_2 \\ \ldots\ldots\ldots\ldots\ldots\ldots\ldots\ldots\ldots \\ a_{m1} x_1 + a_{m2} x_2 + \cdots + a_{mn} x_n = b_m \end{array} \right\} \qquad (2.3)$$

$$\text{i.e.} \quad \boldsymbol{Ax} = \boldsymbol{b}; \boldsymbol{b} > \boldsymbol{0},$$

provided we have a basic feasible solution of the constraints.

Basic solutions of the constraints can be obtained by finding m columns of A which form a non-singular $m \times m$ matrix, B say. If these columns correspond to the variables $x_1, x_2, \ldots, x_m$ the constraints can be solved to express $x_1, \ldots, x_m$ in terms of the b's and the other x's to give

$$\left. \begin{array}{l} x_1 + \qquad\quad + a'_{1m+1} x_{m+1} + a'_{1m+2} x_{m+2} + \cdots + a'_{1n} x_n = b'_1 \\ x_2 + \cdots + a'_{2m+1} x_{m+1} + \cdots\cdots\cdots\cdots + a'_{2n} x_n = b'_2 \\ \ldots\ldots\ldots\ldots\ldots\ldots\ldots\ldots\ldots\ldots\ldots\ldots\ldots\ldots\ldots \\ x_m + a'_{mm+1} x_{m+1} + \qquad\qquad\qquad a'_{mn} x_n = b'_m \end{array} \right\} . \qquad (2.4)$$

If we multiply the constraints of equations (2.4) by $c_1, c_2, \ldots, c_m$ and subtract from z, we eliminate $x_1, x_2, \ldots, x_m$ from z and obtain

$$c'_{m+1} x_{m+1} + c'_{m+2} x_{m+2} + \cdots + c'_n x_n = z - z_0 \qquad (2.5)$$

where $z_0 = \displaystyle\sum_{i=1}^{m} c_i b'_i$.

Of course equations (2.4) and (2.3) represent the same constraints, equations (2.5) and (2.1) the same objective function even though the algebraic forms differ. Equations (2.4) and (2.5) are a **canonical form** for the basis $x_1, x_2, ..., x_m$. If $x_{m+1}, x_{m+2}, ..., x_n$ are put equal to zero, $x_1 = b_1', x_2 = b_2', ..., x_m = b_m', x_{m+1} = 0, ..., x_n = 0$ is a basic solution. If all $b_i' \geqslant 0$, it is a basic feasible solution. The optimum is to be found among such solutions. The Simplex Method provides a systematic procedure for moving from one (feasible) canonical form to another whilst at the same time reducing z.

It should be noted that if the matrix A is partitioned as

$$A = (BR) \tag{2.6}$$

where B is the $m \times m$ matrix of coefficients of $x_1, ..., x_m$ and R is an $m \times (n - m)$ matrix (the rest of A), premultiplication of equations (2.3) by B^{-1} gives equations (2.4). B^{-1} will exist for a basic feasible solution.

For (2.3) is

$$(BR)\, x = b,$$

whence,

$$B^{-1}(BR)\, x = B^{-1}b,$$

i.e.

$$(I_m B^{-1}R)\, x = b', \quad \text{[which is (2.4)]} \tag{2.7}$$

so that

$$b' = B^{-1}b \tag{2.8}$$

and

$$a_j' = B^{-1}a_j, \quad \text{for all columns } j. \tag{2.9}$$

Also

$$c_j' = c_j - \sum_{i=1}^{m} c_i a_{ij}',$$

$$c_j' = c_j - c_B^{\mathsf{T}} a_j' = c_j - c_B^{\mathsf{T}} B^{-1} a_j, \tag{2.10}$$

where $c_B^{\mathsf{T}} = (c_1, c_2, ..., c_m)$ is the row vector of coefficients of the *basic* variables in the original form for z, equation (2.1).

It is not suggested that these ideas are used to search for a basic feasible solution. Such a search method would be very inefficient. Sometimes a basic feasible solution is obvious, and this is the case with the example chosen to illustrate the Simplex Method. It is Example 1 of Section 1.1 which has already been solved graphically.

Example 1
Subject to $x_1, x_2 \geqslant 0$ and

$$3x_1 + 4x_2 \leqslant 1700$$
$$2x_1 + 5x_2 \leqslant 1600$$

minimise

$$-2x_1 - 4x_2 = z.$$

In standard form with non-negative slack variables x_3 and x_4 the constraints and objective function take the form

$$\left. \begin{aligned} 3x_1 + 4x_2 + x_3 \quad &= 1700 \\ 2x_1 + 5x_2 \quad + x_4 &= 1600 \\ -2x_1 - 4x_2 \quad &= z \end{aligned} \right\}. \tag{2.11}$$

Because in this case the $b's$ are positive and the slack variables occur with coefficients $+1$, it is clear that $x_1 = x_2 = 0$, $x_3 = 1700$, $x_4 = 1600$ is a basic feasible solution and equations (2.11) the appropriate canonical form.

z is expressed in terms of the non-basic variables. They have value zero and z has value zero. How can we decrease z? Since x_1 and x_2 must be non-negative any change in their values must be an increase. Because they have *negative* coefficients in the canonical form for z any such increase will decrease z. Rather than increase them both we adopt the simpler strategy of increasing just one of them. Since x_2 has the most negative coefficient we choose x_2. This would appear to be the most rapid way of decreasing z.

However, if we increase x_2, the values of x_3 and x_4 will change since we must satisfy equations (2.11). All values must also remain non-negative. Thus there will be a limit on the amount by which we can increase x_2.

For
$$3x_1 + 4x_2 + x_3 = 1700$$

x_3 is made zero when
$$x_2 = \frac{1700}{4} = 425.$$

For
$$2x_1 + 5x_2 \qquad + x_4 = 1600$$

x_4 is made zero when
$$x_2 = \frac{1600}{5} = 320.$$

Thus we cannot increase x_2 beyond 320 (the minimum of these values) without violating the non-negativity condition on x_4.

The second constraint can be written (on division by 5, the coefficient of x_2) as

$$\tfrac{2}{5} x_1 + x_2 + \qquad \tfrac{1}{5} x_4 = 320.$$

If we subtract 4 times this from the first constraint in equations (2.11) (the '4' is the coefficient of x_2 in this constraint) and -4 times this from the objective function (the '-4' is the coefficient of x_2 in the objective) we eliminate eliminate x_2 from everywhere except the second constraint where it has coefficient 1. The constraints and objective function then appear as

$$\left. \begin{array}{l} \tfrac{7}{5} x_1 \qquad + x_3 - \tfrac{4}{5} x_4 = 420 \\ \tfrac{2}{5} x_1 + x_2 \qquad + \tfrac{1}{5} x_4 = 320 \\ -\tfrac{2}{5} x_1 \qquad + \tfrac{4}{5} x_4 = z + 1280 \end{array} \right\} \qquad (2.12)$$

which is a canonical form for the basis x_2, x_3 which represents a basic feasible solution.

The non-basic variables are x_1 and x_4. They are currently zero. In this case it is only by increasing x_1 that we decrease z. By how much can we increase x_1 whilst still keeping x_2 and x_3 non-negative?

For
$$\tfrac{7}{5} x_1 \qquad + x_3 - \tfrac{4}{5} x_4 = 420$$

x_3 is made zero when
$$x_1 = \frac{420}{\frac{7}{5}} = 300.$$

For
$$\tfrac{2}{5}x_1 + x_2 \quad + \tfrac{1}{5}x_4 = 320$$

x_2 is made zero when
$$x_1 = \frac{320}{\tfrac{2}{5}} = 800.$$

Thus we cannot increase x_1 above 300 (the minimum of these). On division by $\tfrac{7}{5}$ (the coefficient of x_1) the first constraint becomes

$$x_1 + \tfrac{5}{7}x_3 - \tfrac{4}{7}x_4 = 300.$$

We eliminate x_1 from the other constraint and the objective function by subtracting $\tfrac{2}{5}$ and $-\tfrac{2}{5}$ times this from the constraint and objective to obtain a canonical form for the basis x_1, x_2 which is also feasible. It is

$$\left. \begin{array}{l} x_1 \quad + \tfrac{5}{7}x_3 - \tfrac{4}{7}x_4 = 300 \\ x_2 - \tfrac{2}{7}x_3 + \tfrac{3}{7}x_4 = 200 \\ \quad \tfrac{2}{7}x_3 + \tfrac{4}{7}x_4 = z + 1400 \end{array} \right\} . \qquad (2.13)$$

At this stage it will be observed that increases in either of the non-basic variables x_3, x_4, which both have positive coefficients in the form for z, will increase z. Thus it is not possible to reduce z further and we have reached the minimum of z which is -1400 at the basic feasible solution $x_1 = 300$, $x_2 = 200$, $x_3 = x_4 = 0$. If we look back to the geometrical solution of Fig. 1.1 we can see that the successive canonical forms have taken us from O to A to B, the minimum point. The reader is urged to verify that had we chosen to increase x_1 in equations (2.11) the procedure would have taken us from O to C to B.

The calculations of this **iterative process** can be set out conveniently in the so called Simplex Tableaux. These are read just as equations (2.11), (2.12), (2.13) for the constraints. The objective function has been written as $-z - 2x_1 - 4x_2 = 0$; $-z - \tfrac{2}{5}x_1 + \tfrac{4}{5}x_4 = 1280$; $-z + \tfrac{2}{7}x_3 + \tfrac{4}{7}x_4 = 1400$, in the three tableaux which now follow.

Iteration	Basis	Value	x_1	x_2	x_3	x_4
0	x_3	1700	3	4	1	.
	x_4	1600	2	5^*	.	1
	$-z$	0	-2	-4	.	.
1	x_3	420	$\tfrac{7}{5}^*$	.	1	$-\tfrac{4}{5}$
	x_2	320	$\tfrac{2}{5}$	1	.	$\tfrac{1}{5}$
	$-z$	1280	$-\tfrac{2}{5}$	.	.	$\tfrac{4}{5}$
2	x_1	300	1	.	$\tfrac{5}{7}$	$-\tfrac{4}{7}$
	x_2	200	.	1	$-\tfrac{2}{7}$	$\tfrac{3}{7}$
	$-z$	1400	.	.	$\tfrac{2}{7}$	$\tfrac{4}{7}$

At iteration 0 we asterisk the value 5, the coefficient of the variable about to become basic in the critical constraint. At iteration 1 it is the number $\tfrac{7}{5}$ as shown.

Note also the use of dots for zeros which *have to be* zero as opposed to just turning out to be zero.

We can generalise the results just obtained in this particular example. We assume that we have an initial canonical form and that at iteration k have reached the canonical form given by equations (2.4) and (2.5) which in tableau form can be written as follows.

Iteration	Basis	Value	x_1	x_2	x_r	x_m	x_{m+1}	x_s	x_n
	x_1	b_1'	1	.	.	.	.	a_{1m+1}'	a_{1s}'	a_{1n}'
	x_2	b_2'	.	1	.	.	.	a_{2m+1}'	a_{2s}'	a_{2n}'
k				.	.	.	.			
	x_r	b_r'	.	.	1	.	.	a_{rm+1}'	$a_{rs}'^{*}$	a_{rn}'
				.	.	.	.			
	x_m	b_m'	.	.	.	1	.	a_{mm+1}'	a_{ms}'	a_{mn}'
	$-z$	$-z_0'$	.	.	.	.	.	c_{m+1}'	c_s'	c_n'

There are 3 stages in the iterative process.

I Find the Variable to Enter the Basis

The variables $x_{m+1}, ..., x_n$ are non-basic. We find the smallest of $c_{m+1}', ..., c_s',$ $..., c_n'$. Suppose this is c_s'. If c_s' is negative then increasing x_s will decrease z. Thus we adopt the convention, if several of c_j' are negative, of choosing the most negative. This is reasonable but not essential. Any negative c_j' will do. Of course if c_s' is $\geqslant 0$ then z cannot be decreased any more and we have the minimum.

II Find the Variable to Leave the Basis

By how much can we increase x_s without violating the non-negativity conditions on the current basic variables? In the ith constraint, if $a_{is}' > 0$ the largest value that can be given to x_s is b_i'/a_{is}', otherwise x_i will become negative. (N.B. If $a_{is}' \leqslant 0$ the basic variable x_i will increase as x_s increases.) Thus x_s can be increased to the value

$$\max x_s = \min_{\substack{i=1, ..., m \\ a_{is}' > 0}} \left(\frac{b_i'}{a_{is}'} \right). \qquad (2.14)$$

If this minimum occurs in row r, x_r is made zero when x_s is given the value b_r'/a_{rs}'. The other basic variables will stay positive. a_{rs}' is called the pivot element, row r the pivot row, column s the pivot column.

III Construct the New Canonical Form

The new basis is $x_1, x_2, ..., x_s, ..., x_m$ the variable x_r having become non-basic. To construct the new canonical form we make the coefficient of x_s in the pivot row unity by dividing this row by a_{rs}' to form the new pivot row.

Next we eliminate x_s from the other constraints and the objective function. Thus for the ith row ($i \neq r$) in which x_s has coefficient a_{is}' we subtract $a_{is}' \times$ (new pivot row) from the ith row. For the objective function in which x_s has coefficient c_s' (< 0) subtract $c_s' \times$ (new pivot row) from the objective function row.

Thus the canonical form at the next iteration will be:

Iteration	Basis	Value	x_1	x_2			x_r		x_m	x_{m+1}		x_s		x_n
	x_1	b_1^+	1	.		.	a_{1r}^+	.		a_{1m+1}^+		.		a_{1n}^+
	x_2	b_2^+	.	1		.	a_{2r}^+	.		a_{2m+1}^+		.		a_{2n}^+
$k+1$			.	.		.						.		
	x_s	b_r^+	.	.		.	a_{rr}^+	.		a_{rm+1}^+		1		a_{rn}^+
									.					
	x_m	b_m^+	.	.		.	a_{mr}^+	1		a_{mm+1}^+		.		a_{mn}^+
	$-z$	$-z_0^+$	.	.		.	c_r^+			c_{m+1}^+		.		c_n^+

where
$$b_r^+ = b_r'/a_{rs}' \qquad (2.15)$$

$$a_{rj}^+ = a_{rj}'/a_{rs}' \qquad (2.16)$$

$$b_i^+ = b_i' - a_{is}'b_r^+ \qquad (2.17)$$
$$a_{ij}^+ = a_{ij}' - a_{is}'a_{rj}^+ \qquad i \neq r \qquad (2.18)$$

$$c_j^+ = c_j' - c_s'a_{rj}^+ \qquad (2.19)$$

$$z_0^+ = z_0' + c_s'b_r^+ \qquad (2.20)$$

These results (2.15), ..., (2.20) have been recorded for reference purposes. It is not suggested that they are formulae to be remembered. Rather the computations should be done based on the principles contained in step III.

At this stage we have constructed the new canonical form and return to stage I and find the minimum of the c_j^+'s. Eventually we shall find that they are all positive, at which point the minimum of z will have been reached.

Example 2
Subject to $x_1, x_2, x_3, x_4 \geqslant 0$ and

$$2x_1 + 4x_2 + x_3 \qquad = 9$$

$$3x_1 + x_2 \qquad + x_4 = 6$$

minimise $\qquad -6x_1 - 2x_2 \qquad = z.$

This is Example 2 of Section 1.2 in standard form.

The successive tableaux, which the reader should check, follow. The first basis and canonical form is immediately evident.

At iteration 1 the coefficients of the non-basic variables in z are all non-negative. Reference to Fig. 1.3 shows that we have moved from O to C. We have an optimum at C where $x_1 = 2$, $x_3 = 5$, $x_2 = x_4 = 0$ with the minimum for z being -12.

The zero coefficient of x_2 in z shows that we could increase x_2. Although such an increase will not actually decrease z it will not increase it either. This is a case where there is more than one optimum solution.

Iteration	Basis	Value	x_1	x_2	x_3	x_4
0	x_3	9	2	4	1	.
	x_4	6	3^*	1	.	1
	$-z$	0	-6	-2	.	.
1	x_3	5	.	$\frac{10^*}{3}$	1	$-\frac{2}{3}$
	x_1	2	1	$\frac{1}{3}$	.	$\frac{1}{3}$
	$-z$	12	.	0	.	2

The resulting canonical form and corresponding tableau (the pivot element is asterisked) follows.

2	x_2	$\frac{3}{2}$	.	1	$\frac{3}{10}$	$-\frac{2}{10}$
	x_1	$\frac{3}{2}$	1	.	$-\frac{1}{10}$	$\frac{4}{10}$
	$-z$	12	.	.	0	2

This tableau which is again optimal corresponds to the point B in Fig. 1.3. It is clear how multiple optima manifest themselves in the procedure; as zero coefficients in the optimal form for z.

Example 3

Subject to $x_1, x_2, x_3, x_4 \geqslant 0$ and

$$x_1 - x_2 - x_3 \quad\quad = 1$$
$$x_2 \quad\quad + x_4 = 2$$

minimise

$$-x_1 - x_2 \quad\quad\quad = z.$$

It is clear by inspection that $x_1 = 1$, $x_4 = 2$, $x_2 = x_3 = 0$ is a basic feasible solution and that the constraints are in the correct form to apply the method. The objective function contains x_1, one of the basic variables. We can use the first constraint to eliminate x_1 to obtain

$$-2x_2 - x_3 = z + 1.$$

The problem is of course Example 3 of Section 1.2. The first tableau, which corresponds to the point A in Fig. 1.4, appears below.

Iteration	Basis	Value	x_1	x_2	x_3	x_4
0	x_1	1	1	-1	-1	.
	x_4	2	.	1^*	0	1
	$-z$	1	.	-2	-1	.
1	x_1	3	1	.	-1	1
	x_2	2	.	1	0	1
	$-z$	5	.	.	-1	2

The second tableau which is computed in the usual way is also shown. It corresponds to the point B in Fig. 1.4. z can be further decreased by increasing x_3. But now we come to a problem. There are no strictly positive coefficients in the x_3 column in the constraints. Thus, however much we increase x_3 we will never drive a basic variable to zero; indeed x_1 will be increased and x_2 will remain unchanged. We have a case of an unbounded solution which is quite clear from Fig. 1.4. It manifests itself in the Simplex Method through the fact that all $a'_{is} \leqslant 0$.

2.2   Implementing the Simplex Method on the Computer

The computational procedure of the Simplex Method is an iterative process. It would, however, be very tedious if the problem contained several variables and constraints. Many practical problems have tens of variables and constraints (some many more), and it is clear that it will not be reasonable to tackle such problems by way of a hand calculation. But the Simplex Method is a method for the computer. It is no accident that the theoretical developments of linear programming coincided in time with the development of the computer. Without the latter the theory would have had little scope for real applications.

The procedures of the last section can be put in the form of a flow chart for the computation, which can in turn be coded as a computer program.

The iterative routine consists essentially of three steps. First we find the $\min_j c'_j = c'_s (<0)$ to locate the variable to enter the basis. Then the row of the basic variable to leave the basis is found from

$$\max x_s = \min_{\substack{i = 1 \dots m \\ a'_{is} > 0}} \left(\frac{b'_i}{a'_{is}} \right) = \frac{b'_r}{a'_{rs}}. \tag{2.21}$$

This of course is the result (2.14). Finally we find the next canonical form in accordance with equations (2.15)...(2.20).

The program listing follows the flow chart and uses the notation that has been adopted in the text.

```
PROGRAM Simplex (input,output);
CONST
    m=2; n=4;       { No. of constraints and total no. of variables }   {**}
    fwt=8; dpt=2;   { Output format constants for tableau values }      {**}
    fwi=1;          { Output format constant for indices }              {**}
    largevalue = 1.0E20;   smallvalue = 1.0E-10;                        {**}

TYPE   mrange = 1..m;  nrange = 1..n;
    matrix = ARRAY [mrange,nrange] OF real;
    column = ARRAY [mrange] OF REAL;
    baseindex = ARRAY [mrange] OF integer;
    row = ARRAY [nrange] OF real;
    rowboolean = ARRAY [nrange] OF boolean;
```

Flow Chart for Simplex Method given an Initial Canonical Form

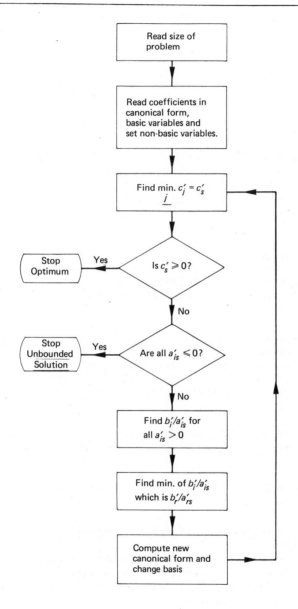

```
VAR
  a : matrix;  { Matrix A in standard form of problem, see (2.3) }
  b : column;  { Vector b in standard form of problem, see (2.3) }
  c : row;     { Coefficients of objective function, see (2.1) }
  basic : baseindex;        { Basic variables at each stage }
  nonbasic : rowboolean;    { Status indicators for variables }
  z0 : real;                { Value of objective function }
  it : integer;             { Iteration counter }
  solution,
  unbounded : boolean;      { Iteration process terminators }
  r, s : integer;           { Row and column of pivot element }


PROCEDURE inputdata;
VAR  i, j : integer;
BEGIN
  FOR i:=1 TO m DO
  BEGIN FOR j:=1 TO n DO read(a[i,j]);  read(b[i]) END;
  FOR j:=1 TO n DO read(c[j]);  read(z0);
  FOR i:=1 TO m DO read(basic[i]); { Initial base }
END; { inputdata }


PROCEDURE initialise;
VAR  i, j : integer;
BEGIN it:=0;  solution:=false;  unbounded:=false;
  FOR j:=1 TO n DO nonbasic[j]:=true;
  FOR i:=1 TO m DO nonbasic[basic[i]]:=false
END; { initialise }


PROCEDURE outputtableau;
VAR  i, j : integer;
BEGIN  writeln; writeln('      ITERATION ', it:2);
  write('      BASE VAR.   ', ' ':fwt-5, 'VALUE');
  FOR j:=1 TO n DO write(' ':fwt-fwi-1, 'X', j:fwi); writeln;
  FOR i:=1 TO m DO
  BEGIN
    write(' ':8-fwi, 'X', basic[i]:fwi, ' ':7, b[i]:fwt:dpt);
    FOR j:=1 TO n DO write(a[i,j]:fwt:dpt); writeln
  END;
  write(' ':7, '-Z', ' ':7, z0:fwt:dpt);
  FOR j:=1 TO n DO write(c[j]:fwt:dpt);  writeln
END; { outputtableau }


PROCEDURE nextbasicvariable (VAR r,s:integer);
VAR  i, j : integer;  min : real;
BEGIN  min:=largevalue;  { Find the variable, s, }
  FOR j:=1 TO n DO        { to enter the basis.   }
    IF nonbasic[j] THEN IF c[j]<min THEN BEGIN min:=c[j]; s:=j END;
  solution := c[s] > -smallvalue;
  IF NOT solution THEN
  BEGIN  unbounded:=true;  i:=1;   { Check that at least one value }
    WHILE unbounded AND (i<=m) DO  { in column s is positive.       }
    BEGIN  unbounded := a[i,s] < smallvalue;  i:=i+1  END;
    IF NOT unbounded THEN
```

```
      BEGIN  min:=largevalue;  { Find the variable, basic[r], }
        FOR i:=1 TO m DO         { to leave the basis.        }
          IF a[i,s] > smallvalue THEN
            IF b[i]/a[i,s] < min THEN BEGIN min:=b[i]/a[i,s]; r:=i END;
        nonbasic[basic[r]]:=true; nonbasic[s]:=false; basic[r]:=s; writeln;
        writeln('    PIVOT IS AT ROW ', r:fwi, ' COL ', s:fwi)
      END
    END
END; { nextbasicvariable }

PROCEDURE transformtableau (r,s:integer);
{ Construct the new canonical form, implementing (2.15) to (2.20) }
VAR  i, j : integer;  pivot, savec : real;  savecol : column;
BEGIN
  FOR i:=1 TO m DO savecol[i]:=a[i,s];  savec:=c[s];  pivot:=a[r,s];
  b[r]:=b[r]/pivot;  {(2.15)}
  FOR j:=1 TO n DO a[r,j]:=a[r,j]/pivot;  {(2.16)}
  FOR i:=1 TO m DO
    IF i<>r THEN
    BEGIN b[i] := b[i] - savecol[i]*b[r];  {(2.17)}
      FOR j:=1 TO n DO a[i,j] := a[i,j] - savecol[i]*a[r,j]  {(2.18)}
    END;
  FOR j:=1 TO n DO c[j] := c[j] - savec*a[r,j];  {(2.19)}
  z0 := z0 - savec*b[r];  {(2.20)}  it := it+1
END; { transformtableau }

BEGIN  { Main Program }
  writeln; writeln('    SIMPLEX METHOD'); writeln;
  inputdata;  initialise;
  REPEAT
    outputtableau;
    nextbasicvariable(r,s);
    IF NOT (solution OR unbounded) THEN transformtableau(r,s)
  UNTIL solution OR unbounded;
  { Output results }  writeln;
  IF unbounded THEN writeln ('    VARIABLE ', s:fwi, ' IS UNBOUNDED')
  ELSE writeln('    MINIMUM AT Z=', -z0:fwt:dpt)
END. { Simplex }
```

A few remarks in addition to those in the program listing may be useful. The minimum c_j is found in the first part of PROCEDURE *nextbasicvariable*, provided it has a negative value. We avoid finding a spurious negative by testing $c[s]$, the minimum c_j, against $-$ *smallvalue* (i.e. -10^{-10}). It must be remembered that the computer only works to finite accuracy. The simplex procedure can be a lengthy and involved calculation and rounding errors can accumulate; in particular values which *should* be zero might be held as say $-1.239\,47 \times 10^{-20}$. We want to avoid such false negatives.

Similar precautions are taken in *nextbasicvariable* to find the row of the variable to leave the basis. We test that a_{is} is genuinely positive. Notice also in this routine, which finds the minimum of b_i/a_{is}, that this means we do not divide by zero. This would cause an execution error.

Boolean variables *solution* and *unbounded* are used to terminate the iterative process.

Intermediate tableaux are output at each iteration by calling the PROCEDURE *outputtableau* from within the REPEAT loop of the Main Program. The position of the pivot is output from the PROCEDURE *nextbasicvariable*. Slight changes to the source code could suppress some of this output if necessary.

Although the output has been formatted (using constants *fwt*, *dpt* and *fwi* which can be changed to suit the accuracies required), it is clear that for large values of n the screen output and hard copy output will overflow the number of columns available, and some modifications will be called for. However, it is good enough as it stands to illustrate the procedure for some fairly small problems with up to 10 variables.

The specimen output given applies to Example 1 of Section 2.1 for which the values in the CONST part of PROGRAM *Simplex* are appropriate and the data file would contain the following values:

$$3 \ 4 \ 1700 \ 2 \ 5 \ 1600 \ -2 \ -4 \ 0 \ 3 \ 4$$

It is clear that the Simplex tableaux of that solution have been reproduced.

The reader is urged to try the program on those other examples for which an initial basic feasible solution is obvious.

```
SIMPLEX METHOD


ITERATION  0
BASE VAR.      VALUE       X1       X2       X3       X4
   X3        1700.00      3.00     4.00     1.00     0.00
   X4        1600.00      2.00     5.00     0.00     1.00
   -Z           0.00     -2.00    -4.00     0.00     0.00


PIVOT IS AT ROW 2 COL 2


ITERATION  1
BASE VAR.      VALUE       X1       X2       X3       X4
   X3         420.00      1.40     0.00     1.00    -0.80
   X2         320.00      0.40     1.00     0.00     0.20
   -Z        1280.00     -0.40     0.00     0.00     0.80


PIVOT IS AT ROW 1 COL 1


ITERATION  2
BASE VAR.      VALUE       X1       X2       X3       X4
   X1         300.00      1.00     0.00     0.71    -0.57
   X2         200.00      0.00     1.00    -0.29     0.43
   -Z        1400.00      0.00     0.00     0.29     0.57

MINIMUM AT Z=-1400.00
```

2.3   Generating an Initial Basic Feasible Solution

The examples chosen to illustrate the Simplex Method were from Section 1.2. They were such that an initial basic feasible solution, and corresponding canonical form, was obvious, or else easy to obtain as in Example 3.

Suppose we try the problem:

Example 1

Subject to $x_1, x_2 \geqslant 0$ and

$$x_1 \qquad \geqslant 10$$
$$x_2 \geqslant 5$$
$$x_1 + x_2 \leqslant 20$$
$$-x_1 + 4x_2 \leqslant 20$$

minimise
$$-3x_1 - 4x_2 = z.$$

This is Example 1 of Section 1.2 and presented no problems to the graphical method. In standard form with non-negative slack variables we can write the constraints as:

$$
\left.
\begin{aligned}
x_1 \quad - x_3 \qquad\qquad\qquad &= 10 \\
x_2 \quad - x_4 \qquad\qquad &= 5 \\
x_1 + x_2 \qquad + x_5 \qquad &= 20 \\
-x_1 + 4x_2 \qquad\qquad + x_6 &= 20 \\
\end{aligned}
\right\} . \qquad (2.22)
$$

minimise
$$-3x_1 - 4x_2 \qquad\qquad = z.$$

However, when we try to use the Simplex Method we do meet a difficulty. We have no obvious basic feasible solution. The basic solution obtained by equating the slack variables to the R.H.S. values is not feasible. This solution is $x_1 = x_2 = 0$, $x_3 = -10$, $x_4 = -5$, $x_5 = 20$, $x_6 = 20$ and x_3 and x_4 are negative. The difficulty arises from the '$\geqslant$' constraints. It would also arise from '$=$' constraints.

One way out of the difficulty is to use the Simplex Method itself to generate a basic feasible solution. We **modify** the first two constraints (there is no problem with the other two) by introducing **artificial variables** (which are non-negative) x_7 and x_8 into the L.H.S. The **modified** constraints are:

$$
\left.
\begin{aligned}
x_1 \quad - x_3 \qquad\qquad + x_7 \quad &= 10 \\
x_2 \quad - x_4 \qquad\qquad + x_8 &= 5 \\
x_1 + x_2 \qquad + x_5 \qquad &= 20 \\
-x_1 + 4x_2 \qquad + x_6 \qquad &= 20 \\
-3x_1 - 4x_2 \qquad\qquad &= z
\end{aligned}
\right\} \qquad (2.23)
$$

for which a basic feasible solution is clear. It is $x_1 = x_2 = x_3 = x_4 = 0$ (the non-basic variables) $x_7 = 10$, $x_8 = 5$, $x_5 = 20$, $x_6 = 20$. We then use the Simplex Method to minimise

$$x_7 + x_8 = w. \qquad (2.24)$$

w is called the **artificial objective function**. Phase I of the problem consists of minimising w.

Provided the constraints (2.22) have a feasible solution, and hence a basic feasible solution (they need not have in all cases), Phase I will end with w having been reduced to zero and with both x_7 and x_8 zero. But when both x_7 and x_8 are zero the modified constraints (2.23) are equivalent to the original constraints (2.22). The basic feasible solution which minimises w can be used as the initial basic feasible solution for the minimisation of z; Phase II of the problem. We just ignore the zero values of x_7 and x_8 from now on.

Of course, to minimise w we must express it in appropriate form, i.e. in terms of the non-basic variables. x_7 and x_8 are of course basic in the first solution of equations (2.23). It is easy to eliminate x_7 and x_8 from w. We just subtract the rows containing them from w. (This idea generalises to other problems.) Thus we obtain

$$- x_1 - x_2 + x_3 + x_4 = w - 15. \tag{2.25}$$

Whilst Phase I proceeds we carry out the appropriate calculations on the z row. It can be treated in the same way as the constraints. In this way when Phase I is completed we shall have z in a form that is appropriate to the current basis. Once we have made w zero we ignore it during Phase II. This also applies to the artificial variables.

The tableaux for Phase I of the problem are as follows.

Iteration	Basis	Value	x_1	x_2	x_3	x_4	x_5	x_6	x_7	x_8
0	x_7	10	1^*	0	-1	0	.	.	1	.
	x_8	5	0	1	0	-1	.	.	.	1
	x_5	20	1	1	0	0	1	.	.	.
	x_6	20	-1	4	0	0	.	1	.	.
	$-z$	0	-3	-4	0	0	.	.	.	.
	$-w$	-15	-1	-1	1	1	.	.	.	.
1	x_1	10	1	0	-1	0	.	.	1	.
	x_8	5	.	1^*	0	-1	.	.	0	1
	x_5	10	.	1	1	0	1	.	-1	.
	x_6	30	.	4	-1	0	.	1	1	.
	$-z$	30	.	-4	-3	0	.	.	.	.
	$-w$	-5	.	-1	0	1	.	.	1	.
2	x_1	10	1	.	-1	0	.	.	1	0
	x_2	5	.	1	0	-1	.	.	0	1
	x_5	5	.	.	1	1	1	.	-1	-1
	x_6	10	.	.	-1	4	.	1	1	-4
	$-z$	50	.	.	-3	-4	.	.	3	4
	$=w$	0	.	.	0	0	.	.	.	.

At this stage we have minimised w, x_7 and x_8 are non-basic and hence zero. Notice, that we could have ignored x_7 as from iteration 1 and certainly we can ignore both x_7 and x_8 from now on. We retained x_7 just to show that in the optimum for w, x_7 and x_8 both appear with coefficient 1 when w is 0 ($-w + x_7 + x_8 = 0$). Because we have kept the form of z appropriate to the basis, we can now proceed with the real business of minimising z. We are now at P in Fig. 1.2. Phase I is now complete and the final tableau less the last two columns and the last row is the first tableau for Phase II. The tableaux for Phase II are as follows:

Iteration	Basis	Value	x_1	x_2	x_3	x_4	x_5	x_6
2	x_1	10	1	.	-1	0	.	.
	x_2	5	.	1	0	-1	.	.
	x_5	5	.	.	1	1	1	.
	x_6	10	.	.	-1	4^*	.	1
	$-z$	50	.	.	-3	-4	.	.
3	x_1	10	1	.	-1	.	.	0
	x_2	$\frac{15}{2}$	.	1	$-\frac{1}{4}$	.	.	$\frac{1}{4}$
	x_5	$\frac{5}{2}$	.	.	$\frac{5}{4}^*$	.	1	$-\frac{1}{4}$
	x_4	$\frac{5}{2}$	.	.	$-\frac{1}{4}$	1	.	$\frac{1}{4}$
	$-z$	60	.	.	-4	.	.	1
4	x_1	12	1	.	.	.	$\frac{4}{5}$	$-\frac{1}{5}$
	x_2	8	.	1	.	.	$\frac{1}{5}$	$\frac{1}{5}$
	x_3	2	.	.	1	.	$\frac{4}{5}$	$-\frac{1}{5}$
	x_4	3	.	.	.	1	$\frac{1}{5}$	$\frac{1}{5}$
	$-z$	68	.	.	.	.	$\frac{16}{5}$	$\frac{1}{5}$

The successive tableaux 2, 3, 4 refer to the points P, Q, R in Fig. 1.2. The last tableau is of course optimal so that the minimum of z is -68 when $x_1 = 12$, $x_2 = 8$, $x_3 = 2$, $x_4 = 3$.

Because the slack variables x_3 and x_4 take on positive values in the optimal solution we see that, at the optimal solution, the constraints in which these variables occur (to take up the slack) are satisfied as **strict inequalities**.

Example 2

(Example 4 of Section 1.2.)
Subject to $x_1, x_2 \geqslant 0$ and

$$x_1 + x_2 \geqslant 10$$
$$3x_1 + 5x_2 \leqslant 15$$

minimise
$$2x_1 + 3x_2 = z.$$

In standard form the constraints become

$$\left.\begin{aligned} x_1 + x_2 - x_3 &= 10 \\ 3x_1 + 5x_2 + x_4 &= 15 \\ 2x_1 + 3x_2 &= z \end{aligned}\right\}. \qquad (2.26)$$

We modify the first constraint by inserting the artificial variable x_5 and minimise $w = x_5$. In canonical form we shall have $-x_1 - x_2 + x_3 = w - 10$. The Simplex tableaux follow.

Iteration	Basis	Value	x_1	x_2	x_3	x_4	x_5
0	x_5	10	1	1	-1	.	1
	x_4	15	3^*	5	0	1	.
	$-z$	0	2	3	0	.	.
	$-w$	-10	-1	-1	1	.	.
1	x_5	5	.	$-\frac{2}{3}$	-1	$-\frac{1}{3}$	1
	x_1	5	1	$\frac{5}{3}$	0	$\frac{1}{3}$	.
	$-z$	-10	.	$-\frac{1}{3}$	0	$-\frac{2}{3}$	.
	$-w$	-5	.	$\frac{2}{3}$	1	$\frac{1}{3}$	.

At this stage w is minimised. All the coefficients in the w row are positive. But w has not been reduced to zero and x_5 is 5. We cannot find a feasible solution to the **unmodified** constraints (2.26). Of course Fig. 1.5 confirms this. Phase I is complete but we cannot start on Phase II because the original constraints do not have a basic feasible solution.

Example 3

A firm requires coal with a phosphorus content no more than 0.03% and no more than 3.25% ash impurity. Three grades of coal A, B, C are available at the prices shown.

Grade	% Phosphorus	% Ash	Cost ($/tonne)
A	0.06	2.0	30
B	0.04	4.0	30
C	0.02	3.0	45

How should these be blended to meet the impurity restrictions at minimum cost? (Question 8 of Exercises 1, in which a hint was given to reduce the problem to a 2-dimensional problem.) We can solve it as it stands using the Simplex Method.

Let 1 tonne of the blend contain x_1, x_2, x_3 tonnes of A, B and C respectively. Then $x_1, x_2, x_3, \geqslant 0$. Further we require

$$x_1 + \quad x_2 + \quad x_3 = 1$$
$$0.06x_1 + 0.04x_2 + 0.02x_3 \leqslant 0.03$$
$$2x_1 + \quad 4x_2 + \quad 3x_3 \leqslant 3.25.$$

Subject to these conditions we require

$$30x_1 + 30x_2 + 45x_3 = z \text{ to be a minimum.}$$

If we add slack variables to the second and third constraints (first multiply constraint 2 by 100) we have the problem in standard form.

For non-negative x_i $(i = 1, ..., 5)$ subject to

$$x_1 + x_2 + x_3 \qquad\qquad = 1$$
$$6x_1 + 4x_2 + 2x_3 + x_4 \qquad = 3$$
$$2x_1 + 4x_2 + 3x_3 \qquad + x_5 = 3\tfrac{1}{4}$$

minimise $\qquad\qquad 30x_1 + 30x_2 + 45x_3 = z.$

A basic feasible solution is not obvious. We treat the first 'equation' constraint as we did the inequality constraints in the earlier examples. We add an artificial variable x_6. Then we minimise $w = x_6$ and use the resulting optimal solution as the starting point for minimising z.

Thus the modified constraints become

$$x_1 + x_2 + x_3 \qquad\qquad + x_6 = 1$$
$$6x_1 + 4x_2 + 2x_3 + x_4 \qquad\qquad = 3$$
$$2x_1 + 4x_2 + 3x_3 \qquad\qquad + x_5 \quad = 3\tfrac{1}{4}$$
$$30x_1 + 30x_2 + 45x_3 \qquad\qquad = z$$

minimise $\qquad\qquad\qquad\qquad x_6 = w,$

for which a basic feasible solution is $x_1 = x_2 = x_3 = 0$, $x_6 = 1$, $x_4 = 3$, $x_5 = 3\tfrac{1}{4}$. We express w in terms of the non-basic variables by eliminating x_6. We just subtract the first constraint (the only one containing x_6, and with coefficient 1) from w to obtain

$$- x_1 - x_2 - x_3 \qquad\qquad = w - 1.$$

The calculations of the successive tableaux are shown on the next page.

Phase I ends after iteration 1, from which point we ignore w and x_6. The minimum value of z is $\$38\tfrac{3}{4}$ when $x_1 = \tfrac{1}{12}$, $x_2 = \tfrac{1}{3}$, $x_3 = \tfrac{7}{12}$.

2.4 The Full Simplex Method

We have seen that L.P. problems can arise with the constraints in a variety of forms. A basic feasible solution is not always apparent. Thus any computer routine should be capable itself of setting up the slack and artificial variables as required and generating the first basic feasible solution.

We suppose that our problem has M constraints in the N (original) variables. We suppose that there are GC '$\geqslant$', EC '$=$' and LC '$\leqslant$' constraints and that they are *arranged in this order* in the problem. We suppose that the R.H.S. values of the constraints are positive. There is no loss of generality in this. The original problem has

Iteration	Basis	Value	x_1	x_2	x_3	x_4	x_5	x_6
0	x_6	1	1	1	1^*	.	.	1
	x_4	3	6	4	2	1	.	.
	x_5	$3\frac{1}{4}$	2	4	3	.	1	.
	$-z$	0	30	30	45	.	.	.
	$-w$	-1	-1	-1	-1	.	.	.
1	x_3	1	1	1	1	.	.	1
	x_4	1	4	2	.	1	.	-2
	x_5	$\frac{1}{4}$	-1	1^*	.	.	1	-3
	$-z$	-45	-15	-15	.	.	.	-45
	$-w$	0	0	0	0			
2	x_3	$\frac{3}{4}$	2	.	1	.	-1	
	x_4	$\frac{1}{2}$	6^*	.	.	1	-2	
	x_2	$\frac{1}{4}$	-1	1	.	.	.	
	$-z$	$-41\frac{1}{4}$	-30	.	.	.	15	
3	x_3	$\frac{7}{12}$	.	.	1	.	$-\frac{1}{3}$	$-\frac{1}{3}$
	x_1	$\frac{1}{12}$	1	.	.	.	$\frac{1}{6}$	$-\frac{1}{3}$
	x_2	$\frac{1}{3}$	.	1	.	.	$\frac{1}{6}$	$\frac{2}{3}$
	$-z$	$-38\frac{3}{4}$	.	.	.		5	5

the form:

$$x_i \geq 0, \quad i = 1, \ldots, N$$

$$
\begin{array}{l}
a_{11}x_1 + \cdots + a_{1N}x_N \geq b_1 \\
\hphantom{a_{11}x_1 + \cdots + a_{1N}x_N} \geq b_{GC} \\
\hphantom{a_{11}} = b_{GC+1} \\
\hphantom{a_{11}} = b_{GC+EC} \\
\hphantom{a_{11}} \leq b_{GC+EC+1} \\
a_{M1}x_1 + \cdots + a_{MN}x_N \leq b_M
\end{array}
\quad
\begin{array}{l}
\left.\rule{0pt}{22pt}\right\}\ \text{GC rows} \\
\left.\rule{0pt}{22pt}\right\}\ \text{EC rows} \\
\left.\rule{0pt}{22pt}\right\}\ \text{LC rows}
\end{array}
\ \left.\rule{0pt}{66pt}\right\}\ \text{M rows}
$$

$c_1 x_1 + \cdots + c_N x_N = z$ is to be minimised.

Slack variables $x_{N+1} \ldots x_{N+GC}$ with coefficient -1 are put in the first GC constraints. Slack variables $x_{N+GC+1} \ldots x_{N+GC+LC}$ with coefficient $+1$ are put in the last LC constraints. Artificial variables $x_{N+GC+LC+1} \ldots x_{N+GC+LC+EC}$ with coefficients $+1$ are put into the first GC + EC constraints. These variables along with the last LC slack variables form the initial basic feasible solution. Their values are the corresponding b's.

The artificial objective function w consists of the sum of the artificial variables. When expressed in terms of the non-basic variables it takes the form

$$d_1 x_1 + \cdots + d_N x_N + \cdots + d_{N+GC+LC} x_{N+GC+LC} = w + w_0$$

where d_j consists of the negative value of the sum over the first GC + EC rows of the coefficient of the variable x_j in the enhanced matrix of the constraints. w_0 is the negative of the sum of the corresponding b's. w_0 and the d's appear in row M + 2 of the enhanced tableau.

Thus for the problem in which all variables are non-negative and GC = 2, EC = 1, LC = 1 (i.e. M = 4, GC + LC = 3) and N = 3

$$2x_1 + 3x_2 + x_3 \geqslant 6$$
$$x_1 + 5x_2 + 6x_3 \geqslant 9$$
$$x_1 + x_2 + x_3 = 4$$
$$-x_1 + x_3 \leqslant 2$$
$$-x_1 - 2x_2 - 3x_3 = z \text{ is to be minimised}$$

we want to set up the first tableau.

Iteration	Basis	Value	x_1	x_2	x_3	x_4	x_5	x_6	x_7	x_8	x_9
0	x_7	6	2	3	1	−1	0	.	1	.	.
	x_8	9	1	5	6	0	−1	.	.	1	.
	x_9	4	1	1	1	0	0	.	.	.	1
	x_6	2	−1	0	1	0	0	1	.	.	.
	$-z$	0	−1	−2	−3	0	0	.	.	.	.
	$-w$	−19	−4	−9	−8	1	1	.	.	.	.

We can use parts of the program of Section 2.2 to carry out the computations and these have been included in the PROCEDURE *Simplex* of the following PROGRAM *FullSimplex*. We need to know whether we are in Phase I or Phase II and this is indicated through the procedure parameter p. When we are in Phase I the objective is in row d and z is carried along just like any other constraint. When we are in Phase II the objective function is in row c and we ignore columns of the artificial variables. When at the end of Phase I, we have minimised w, we have to check that we have reduced it to zero (apart from a rounding error) and then call *Simplex* with parameter *Phase II*. The program listing follows.

```
PROGRAM FullSimplex (input,output);
CONST
   nvar=2; m=4;          { No. of variables and constraints }     {**}
   ncols=8;              { Maximum no. of columns in tableau }    {**}
   fwt=7;  dpt=2;        { Output format constants for tableau values }  {**}
   fwi=1;               { Output format constant for indices }    {**}
   largevalue = 1.0E20;  smallvalue=1.0E-10;                      {**}

TYPE   mrange = 1..m;  ncolsrange = 1..ncols;
   matrix = ARRAY [mrange,ncolsrange] OF real;
   column = ARRAY [mrange] OF real;
   baseindex = ARRAY [mrange] OF integer;
   row = ARRAY [ncolsrange] OF real;
   rowboolean = ARRAY [ncolsrange] OF boolean;
   phase = (PhaseI, PhaseII);
```

```
VAR
  a : matrix;   { Matrix A in standard form of problem, see (2.3) }
  b : column;   { Vector b in standard form of problem, see (2.3) }
  c : row;      { Coefficients of objective function, see (2.1) }
  d : row;      { Coefficients of artificial objective function }
  basic : baseindex;       { Basic variables at each stage }
  nonbasic : rowboolean;   { Status indicators for variables }
  w0, z0 : real;           { Values of objective functions }
  it : integer;            { Iteration counter }
  solution, OK : boolean;  { Iteration process terminators }
  r, s : integer;          { Row and column of pivot element }
  GC, EC, LC : integer;    { No. of ´>=´, ´=´ and ´<=´ constraints }
  nl, n2, GCplusLC, GCplusEC : integer;
  printon : boolean;  i : integer;   slack : row;



PROCEDURE inputdata;
VAR  i, j, k : integer;
BEGIN  read(k); printon:=k>0; read(GC,EC,LC);
  FOR i:=1 TO m DO
  BEGIN FOR j:=1 TO nvar DO read(a[i,j]);  read(b[i]) END;
  FOR j:=1 TO nvar DO read(c[j])
END; { inputdata }



PROCEDURE initialise;
VAR  i, j : integer;
BEGIN  it:=0; z0:=0.0; OK:=true; GCplusLC:=GC+LC; GCplusEC:=GC+EC;
  nl := nvar + GCplusLC + GCplusEC;   n2 := nvar + GCplusLC;
  FOR j:= nvar+1 TO nl DO
  BEGIN FOR i:=1 TO m DO a[i,j]:=0.0; c[j]:=0.0 END
END; { initialise }



PROCEDURE completetableau;
VAR  i, j : integer;  sum : real;
BEGIN  FOR i:=1 TO GC DO a[i,nvar+i] := -1.0;
  FOR i:=1 TO LC DO a[GCplusEC+i,nvar+GC+i] := 1.0;
  FOR i:=1 TO GCplusEC DO a[i,nvar+GCplusLC+i] := 1.0;
  { Compute initial base }
  FOR j:=1 TO GCplusEC DO basic[j] := nvar + GCplusLC + j;
  FOR j:=1 TO LC DO basic[GCplusEC+j] := nvar + GC + j;
  FOR j:=1 TO nl DO nonbasic[j]:=true;
  FOR i:=1 TO m DO nonbasic[basic[i]]:=false;
  {Compute d-values and w0 }
  FOR j:=1 TO n2 DO
  BEGIN  sum:=0.0;
    FOR i:=1 TO GCplusEC DO sum := sum + a[i,j];  d[j]:=-sum
  END;  sum:=0.0;
  FOR j:=n2+1 TO nl DO d[j]:=0.0;
  FOR i:=1 TO GCplusEC DO sum := sum + b[i];  w0:=-sum
END; { completetableau }
```

```
PROCEDURE outputtableau (p:phase);
VAR  i, j, n : integer;
BEGIN  IF p=PhaseI THEN n:=n1 ELSE n:=n2;
  writeln; writeln('    ITERATION ', it:2);
  write('     BASE VAR.   ', '  ':fwt-5, 'VALUE');
  FOR j:=1 TO n DO write(' ':fwt-fwi-1, 'X', j:fwi); writeln;
  FOR i:=1 TO m DO
  BEGIN
    write(' ':8-fwi, 'X', basic[i]:fwi, ' ':7, b[i]:fwt:dpt);
    FOR j:=1 TO n DO write(a[i,j]:fwt:dpt); writeln
  END;
  write(' ':7, '-Z', ' ':7, z0:fwt:dpt);
  FOR j:=1 TO n DO write(c[j]:fwt:dpt);  writeln;
  IF p=PhaseI THEN
  BEGIN  write(' ':7, '-W', ' ':7, w0:fwt:dpt);
    FOR j:=1 TO n DO write(d[j]:fwt:dpt);  writeln
  END
END; { outputtableau }


PROCEDURE Simplex (p:phase);
VAR  n : integer;  unbounded : boolean;


  PROCEDURE nextbasicvariable (VAR r,s:integer; x:row);
  VAR  i, j : integer;  min : real;
  BEGIN  min:=largevalue;  { Find the variable, s, }
    FOR j:=1 TO n DO        { to enter the basis.  }
      IF nonbasic[j] THEN IF x[j]<min THEN BEGIN min:=x[j]; s:=j END;
    solution := x[s] > -smallvalue;
    IF NOT solution THEN
    BEGIN  unbounded:=true; i:=1;  { Check that at least one value }
      WHILE unbounded AND (i<=m) DO { in column s is positive.     }
      BEGIN  unbounded := a[i,s] < smallvalue;  i:=i+1  END;
      IF NOT unbounded THEN
      BEGIN  min:=largevalue;  { Find the variable, basic[r], }
        FOR i:=1 TO m DO        { to leave the basis.       }
          IF a[i,s] > smallvalue THEN
            IF b[i]/a[i,s] < min THEN BEGIN min:=b[i]/a[i,s]; r:=i END;
            nonbasic[basic[r]]:=true; nonbasic[s]:=false; basic[r]:=s; writeln;
            writeln('    PIVOT IS AT ROW ', r:fwi, ' COL ', s:fwi)
      END
    END
  END; { nextbasicvariable }


  PROCEDURE transformtableau (r,s:integer; VAR x:row; VAR x0:real);
  { Construct the new canonical form, implementing (2.15) to (2.20) }
  VAR  i, j : integer;  pivot, savec, savex : real;  savecol : column;
  BEGIN
    FOR i:=1 TO m DO savecol[i]:=a[i,s];  savex:=x[s];  pivot:=a[r,s];
    b[r]:=b[r]/pivot;  {(2.15)}
    FOR j:=1 TO n DO a[r,j]:=a[r,j]/pivot;  {(2.16)}
    FOR i:=1 TO m DO
      IF i<>r THEN
```

```
          BEGIN b[i] := b[i] - savecol[i]*b[r];  {(2.17)}
             FOR j:=1 TO n DO a[i,j] := a[i,j] - savecol[i]*a[r,j]  {(2.18)}
          END;
        FOR j:=1 TO n DO x[j] := x[j] - savex*a[r,j];   {(2.19)}
        x0 := x0 - savex*b[r];  {(2.20)}  it := it+1;
        IF p=PhaseI THEN
        BEGIN  savec:=c[s]; FOR j:=1 TO n DO c[j] := c[j] - savec*a[r,j];
          z0 := z0 - savec*b[r]
        END
    END; { transformtableau }




BEGIN   { Simplex }
   solution:=false; unbounded:=false;
   IF p=PhaseI THEN n:=n1 ELSE n:=n2;  { Determine current tableau size }
   REPEAT
     IF printon THEN outputtableau(p);
     CASE p OF
       PhaseI  : nextbasicvariable(r,s,d);
       PhaseII : nextbasicvariable(r,s,c)
     END;
     IF NOT (solution OR unbounded) THEN
       CASE p OF
         PhaseI  : transformtableau(r,s,d,w0);
         PhaseII : transformtableau(r,s,c,z0)
       END
   UNTIL solution OR unbounded;
   IF unbounded THEN writeln('    UNBOUNDED')
END; { Simplex }




BEGIN   { Main Program }
   writeln; writeln('    FULL SIMPLEX METHOD'); writeln;
   inputdata;  initialise;  completetableau;
   IF GCplusEC=0 THEN writeln('    THERE IS NO PHASE I')
   ELSE  { Perform Phase I }
   BEGIN  writeln('    PHASE I');
     Simplex(PhaseI);  writeln;
     IF (abs(w0)>smallvalue) OR (NOT solution) THEN
     BEGIN  OK:=false;  writeln('    PHASE I NOT COMPLETED');
          writeln('    SUM OF ARTIFICIALS ', w0:fwt,dpt)
     END
     ELSE
     BEGIN  writeln; writeln('    PHASE I SUCCESSFUL');  writeln;
       writeln('    REDUCED TABLEAU FOR PHASE II')
     END
   END;
   IF OK THEN  { Perform Phase II }
   BEGIN  Simplex(PhaseII);  writeln;
     IF NOT solution THEN writeln('    PHASE II NOT COMPLETED')
     ELSE
```

```
BEGIN  { Output final details }
  writeln; writeln('     FINAL SOLUTION'); writeln;
  writeln('     MINIMUM OF Z = ', -z0:fwt:dpt); writeln;
  writeln('     CONSTRAINT    BASIS      VALUE      STATE          SLACK');
  FOR i:=1 TO m DO slack[basic[i]] := b[i];
  FOR i:=1 TO m DO { For each constraint }
  BEGIN  write(i:10, basic[i]:10, ' ':12-fwt, b[i]:fwt:dpt, ' ':5);
    IF (i<=GC) OR (i>GCplusEC) THEN
      IF nonbasic[nvar+i] THEN writeln('BINDING', 0.0:10:dpt)
      ELSE  writeln('SLACK', ' ':12-fwt, slack[nvar+i]:fwt:dpt)
    ELSE  writeln('EQUATION      NONE')
  END
 END
END
END. { FullSimplex }
```

The Main Program part of *FullSimplex* calls the procedures to input the data, initialise the variables and complete the computation of tableau values. It then proceeds to call *Simplex* for Phase I (if required) and, on satisfactory completion, calls *Simplex* for Phase II. Non-completion of either phase is reported. Intermediate output can be suppressed by making the first data item non-positive.

Example 1

Find non-negative x_1, x_2 such that

$$x_1 \geqslant 10$$
$$x_2 \geqslant 5$$
$$x_1 + x_2 \leqslant 20$$
$$-x_1 + 4x_2 \leqslant 20$$

and

$$-3x_1 - 4x_2 = z \text{ is minimised.}$$

This is Example 1 of Section 2.3. GC = 2, EC = 0, LC = 2, N = 2. The values in the CONST section of *FullSimplex* are appropriate for this example (note that *ncols* = N + GC + LC + GC + EC = 8 here) and the data file to give the output below would contain the following values:

$$1\ 2\ 0\ 2\ 1\ 0\ 10\ 0\ 1\ 5\ 1\ 1\ 20\ -1\ 4\ 20\ -3\ -4.$$

The output shown clearly reproduces the tableaux which were calculated before.

```
FULL SIMPLEX METHOD

PHASE I

ITERATION  0
```

BASE VAR.	VALUE	X1	X2	X3	X4	X5	X6	X7	X8
X7	10.00	1.00	0.00	-1.00	0.00	0.00	0.00	1.00	0.00
X8	5.00	0.00	1.00	0.00	-1.00	0.00	0.00	0.00	1.00
X5	20.00	1.00	1.00	0.00	0.00	1.00	0.00	0.00	0.00
X6	20.00	-1.00	4.00	0.00	0.00	0.00	1.00	0.00	0.00
-Z	0.00	-3.00	-4.00	0.00	0.00	0.00	0.00	0.00	0.00
-W	-15.00	-1.00	-1.00	1.00	1.00	0.00	0.00	0.00	0.00

PIVOT IS AT ROW 1 COL 1

ITERATION   1

BASE VAR.	VALUE	X1	X2	X3	X4	X5	X6	X7	X8
X1	10.00	1.00	0.00	-1.00	0.00	0.00	0.00	1.00	0.00
X8	5.00	0.00	1.00	0.00	-1.00	0.00	0.00	0.00	1.00
X5	10.00	0.00	1.00	1.00	0.00	1.00	0.00	-1.00	0.00
X6	30.00	0.00	4.00	-1.00	0.00	0.00	1.00	1.00	0.00
-Z	30.00	0.00	-4.00	-3.00	0.00	0.00	0.00	3.00	0.00
-W	-5.00	0.00	-1.00	0.00	1.00	0.00	0.00	1.00	0.00

PIVOT IS AT ROW 2 COL 2

ITERATION   2

BASE VAR.	VALUE	X1	X2	X3	X4	X5	X6	X7	X8
X1	10.00	1.00	0.00	-1.00	0.00	0.00	0.00	1.00	0.00
X2	5.00	0.00	1.00	0.00	-1.00	0.00	0.00	0.00	1.00
X5	5.00	0.00	0.00	1.00	1.00	1.00	0.00	-1.00	-1.00
X6	10.00	0.00	0.00	-1.00	4.00	0.00	1.00	1.00	-4.00
-Z	50.00	0.00	0.00	-3.00	-4.00	0.00	0.00	3.00	4.00
-W	0.00	0.00	0.00	0.00	0.00	0.00	0.00	1.00	1.00

PHASE I SUCCESSFUL

REDUCED TABLEAU FOR PHASE II

ITERATION   2

BASE VAR.	VALUE	X1	X2	X3	X4	X5	X6
X1	10.00	1.00	0.00	-1.00	0.00	0.00	0.00
X2	5.00	0.00	1.00	0.00	-1.00	0.00	0.00
X5	5.00	0.00	0.00	1.00	1.00	1.00	0.00
X6	10.00	0.00	0.00	-1.00	4.00	0.00	1.00
-Z	50.00	0.00	0.00	-3.00	-4.00	0.00	0.00

PIVOT IS AT ROW 4 COL 4

ITERATION   3

BASE VAR.	VALUE	X1	X2	X3	X4	X5	X6
X1	10.00	1.00	0.00	-1.00	0.00	0.00	0.00
X2	7.50	0.00	1.00	-0.25	0.00	0.00	0.25
X5	2.50	0.00	0.00	1.25	0.00	1.00	-0.25
X4	2.50	0.00	0.00	-0.25	1.00	0.00	0.25
-Z	60.00	0.00	0.00	-4.00	0.00	0.00	1.00

PIVOT IS AT ROW 3 COL 3

ITERATION   4

BASE VAR.	VALUE	X1	X2	X3	X4	X5	X6
X1	12.00	1.00	0.00	0.00	0.00	0.80	-0.20
X2	8.00	0.00	1.00	0.00	0.00	0.20	0.20
X3	2.00	0.00	0.00	1.00	0.00	0.80	-0.20
X4	3.00	0.00	0.00	0.00	1.00	0.20	0.20
-Z	68.00	0.00	0.00	0.00	0.00	3.20	0.20

FINAL SOLUTION

MINIMUM OF Z = -68.00

CONSTRAINT	BASIS	VALUE	STATE	SLACK
1	1	12.00	SLACK	2.00
2	2	8.00	SLACK	3.00
3	3	2.00	BINDING	0.00
4	4	3.00	BINDING	0.00

2.5 The Problem of Degeneracy

In the examples considered so far all the basic variables have been non-zero. Clearly all the non-basic variables are zero (by definition). However, there is no reason why one or more of the basic variables should not turn out to be zero. In such a case the basis is called **degenerate** and difficulties *may* arise.

Suppose that one of the current basic variables x_r is zero ($b_r' = 0$). If x_s is the basic variable about to enter the basis then if a_{rs}' is positive, a_{rs}' will be the pivot; b_r'/a_{rs}' will be zero and hence the

$$\min_{\substack{i \\ a_{is}' > 0}} \frac{b_i'}{a_{is}'} = 0.$$

Hence x_s will enter the basis with value 0 and the new basis will also be degenerate. The value of z will be unchanged. If it were then possible for the next iteration to reverse this step and replace x_s by x_r we should be in a closed loop and the iterations would fail to make a further reduction in z.

Example 1

Find non-negative x_1, x_2 such that

$$2x_1 - x_2 \leqslant 4$$
$$x_1 - 2x_2 \leqslant 2$$
$$x_1 + x_2 \leqslant 5$$

and
$$-3x_1 + x_2 = z \text{ is minimised.}$$

The graphical solution (Fig. 2.1) shows the minimum point to be $(3, 2)$ where $z = -7$. The degeneracy arises because three of the lines of the constraints pass through the point $(2, 0)$. Normally a vertex arises from the intersection of just two lines (in two dimensions). Here we have three lines intersecting at a vertex.

If we use the Simplex Method the first tableau is

Iteration	Basis	Value	x_1	x_2	x_3	x_4	x_5
0	x_3	4	2*	-1	1	.	.
	x_4	2	1 (*)	-2	.	1	.
	x_5	5	1	1	.	.	1
	$-z$	0	-3	1	.	.	.

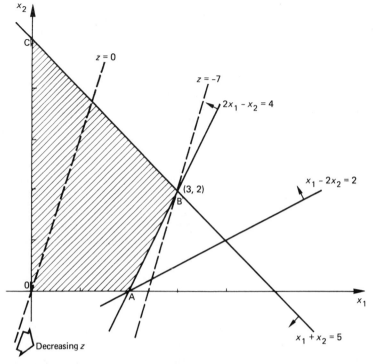

Figure 2.1

x_1 enters the basis. But which variable leaves the basis? There is a *tie* for the minimum; $\frac{4}{2} = \frac{2}{1}$. Should it be x_3 or x_4? Whichever we choose the other will become zero at the next iteration and the basis will be degenerate. Suppose we choose x_3. The other tableaux follow readily.

Iteration	Basis	Value	x_1	x_2	x_3	x_4	x_5
1	x_1	2	1	$-\frac{1}{2}$	$\frac{1}{2}$	.	.
	x_4	0	.	$-\frac{3}{2}$	$-\frac{1}{2}$	1	.
	x_5	3	.	$\frac{3}{2}*$	$-\frac{1}{2}$	.	1
	$-z$	6	.	$-\frac{1}{2}$	$\frac{3}{2}$	.	.
2	x_1	3	1	.	$\frac{1}{3}$	.	$\frac{1}{3}$
	x_4	3	.	.	-1	1	1
	x_2	2	.	1	$-\frac{1}{3}$	.	$\frac{2}{3}$
	$-z$	7	.	.	$\frac{4}{3}$	.	$\frac{1}{3}$

Thus the minimum of z is -7 when $x_1 = 3$, $x_2 = 2$.

Suppose at iteration zero that we choose x_4 to leave the basis. The tableaux are then as given on the next page. Again the final solution is the same as before.

Iteration	Basis	Value	x_1	x_2	x_3	x_4	x_5
1	x_3	0	.	3^*	1	-2	.
	x_1	2	1	-2	.	1	.
	x_5	3	.	3	.	-1	1
	$-z$	6	.	-5	.	3	.
2	x_2	0	.	1	$\frac{1}{3}$	$-\frac{2}{3}$	.
	x_1	2	1	.	$\frac{2}{3}$	$-\frac{1}{3}$	.
	x_5	3	.	.	-1	1^*	1
	$-z$	6	.	.	$\frac{5}{3}$	$-\frac{1}{3}$	.
3	x_2	2	.	1	$-\frac{1}{3}$	.	$\frac{2}{3}$
	x_1	3	1	.	$\frac{1}{3}$	.	$\frac{1}{3}$
	x_4	3	.	.	-1	1	1
	$-z$	7	.	.	$\frac{4}{3}$	.	$\frac{1}{3}$

In this case the degenerate basis at iteration 1 is changed to a second degenerate basis at iteration 2. z is not changed. Then the process moves to the minimum at B. One way to remove the degeneracy would be to perturb the constraints through A by small amounts. Dantzig suggested replacing them by

$$2x_1 - x_2 \leqslant 4 + \varepsilon$$
$$x_1 - 2x_2 \leqslant 2 + \varepsilon^2$$

where ε is a small quantity. This will remove the degeneracy since the perturbed constraints will not be concurrent at A. Rather the region around A will be as in Fig. 2.2.

We then solve the non-degenerate problem. The final solution to this will involve ε. We then let ε become zero. In this way we could sidestep degeneracy. It may all seem rather unnecessarily complicated but readers are urged not to worry at this stage. We shall take up the point in more detail later on.

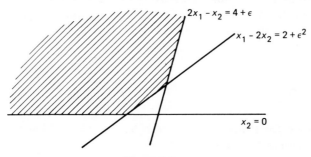

Figure 2.2

If there is no degeneracy the Simplex Method will solve the general L.P. problem with m constraints in n variables (provided of course that it has a solution) in a *finite* number of iterations.

This is easy to establish. At any iteration we have seen from equation (2.20) that the value of the objective function changes from z_0' say to $z_0' + c_s'b_r^+$ where c_s' is the coefficient of the variable which has just entered the basis in the previous objective function, and b_r^+ is the value it assumes. Now c_s' is strictly negative and b_r^+ is positive. Thus z is *decreased* at each stage. Thus we can never return to a previous set of basic variables which would imply a return to a previous value of the objective. But since there are at most $\binom{n}{m}$ basic feasible solutions we must find the minimum in at most $\binom{n}{m}$ iterations.

Notice that the argument will break down if b_r^+ could be zero. It would then be possible to return to a previous basis.

This phenomenon is called **cycling**. It is rare and can probably be avoided in hand calculations by making the 'right choices' when ties occur. Computer programs should be able to handle this problem. We shall return to this point later.

Example 2

This example illustrates the phenomenon of cycling. It was constructed by E. M. L. Beale.

Find non-negative x_1, x_2, x_3, x_4 such that

$$\tfrac{1}{4} x_1 - 8x_2 - x_3 + 9x_4 \leqslant 0$$
$$\tfrac{1}{2} x_1 - 12x_2 - \tfrac{1}{2} x_3 + 3x_4 \leqslant 0$$
$$x_3 \qquad \leqslant 1$$

and
$$-\tfrac{3}{4} x_1 + 20x_2 - \tfrac{1}{2} x_3 + 6x_4 = z \text{ is minimised.}$$

The computer solution (fortuitously) illustrates the cycling. (This shows that to make the program totally robust some changes are needed.) At iteration zero there is a choice of variable to become non-basic, x_5 or x_6. x_5 has been chosen. At iteration 2, x_1 rather than x_2 has been chosen. At iteration 4, x_3 rather than x_4 has been chosen. The tableau at iteration 6 is identical with the initial tableau. The computer will go on repeating the calculations without reducing z. In a hand calculation we should use our discretion and modify the choice made where there are 'ties'. The reader is urged to do this.

It is interesting to note that by changing the order of the first and second constraint the problem is avoided on the computer. The output for the same problem but with the constraints put in the order

$$\tfrac{1}{2} x_1 - 12x_2 - \tfrac{1}{2} x_3 + 3x_4 \leqslant 0$$
$$\tfrac{1}{4} x_1 - 8x_2 - x_3 + 9x_4 \leqslant 0$$
$$x_3 \qquad \leqslant 1$$

is given.

The minimum for z arises when $x_1 = 1$, $x_3 = 1$, $x_6 = 0.75$ and all other x_i's are zero. The minimum for z is -1.25.

FULL SIMPLEX METHOD

THERE IS NO PHASE I

ITERATION   0

BASE VAR.	VALUE	X1	X2	X3	X4	X5	X6	X7
X5	0.00	0.25	-8.00	-1.00	9.00	1.00	0.00	0.00
X6	0.00	0.50	-12.00	-0.50	3.00	0.00	1.00	0.00
X7	1.00	0.00	0.00	1.00	0.00	0.00	0.00	1.00
-Z	0.00	-0.75	20.00	-0.50	6.00	0.00	0.00	0.00

PIVOT IS AT ROW 1 COL 1

ITERATION   1

BASE VAR.	VALUE	X1	X2	X3	X4	X5	X6	X7
X1	0.00	1.00	-32.00	-4.00	36.00	4.00	0.00	0.00
X6	0.00	0.00	4.00	1.50	-15.00	-2.00	1.00	0.00
X7	1.00	0.00	0.00	1.00	0.00	0.00	0.00	1.00
-Z	0.00	0.00	-4.00	-3.50	33.00	3.00	0.00	0.00

PIVOT IS AT ROW 2 COL 2

ITERATION   2

BASE VAR.	VALUE	X1	X2	X3	X4	X5	X6	X7
X1	0.00	1.00	0.00	8.00	-84.00	-12.00	8.00	0.00
X2	0.00	0.00	1.00	0.37	-3.75	-0.50	0.25	0.00
X7	1.00	0.00	0.00	1.00	0.00	0.00	0.00	1.00
-Z	0.00	0.00	0.00	-2.00	18.00	1.00	1.00	0.00

PIVOT IS AT ROW 1 COL 3

ITERATION   3

BASE VAR.	VALUE	X1	X2	X3	X4	X5	X6	X7
X3	0.00	0.12	0.00	1.00	-10.50	-1.50	1.00	0.00
X2	0.00	-0.05	1.00	0.00	0.19	0.06	-0.12	0.00
X7	1.00	-0.12	0.00	0.00	10.50	1.50	-1.00	1.00
-Z	0.00	0.25	0.00	0.00	-3.00	-2.00	3.00	0.00

PIVOT IS AT ROW 2 COL 4

ITERATION   4

BASE VAR.	VALUE	X1	X2	X3	X4	X5	X6	X7
X3	0.00	-2.50	56.00	1.00	0.00	2.00	-6.00	0.00
X4	0.00	-0.25	5.33	0.00	1.00	0.33	-0.67	0.00
X7	1.00	2.50	-56.00	0.00	0.00	-2.00	6.00	1.00
-Z	0.00	-0.50	16.00	0.00	0.00	-1.00	1.00	0.00

PIVOT IS AT ROW 1 COL 5

ITERATION   5

BASE VAR.	VALUE	X1	X2	X3	X4	X5	X6	X7
X5	0.00	-1.25	28.00	0.50	0.00	1.00	-3.00	0.00
X4	0.00	0.17	-4.00	-0.17	1.00	0.00	0.33	0.00
X7	1.00	0.00	0.00	1.00	0.00	0.00	0.00	1.00
-Z	0.00	-1.75	44.00	0.50	0.00	0.00	-2.00	0.00

PIVOT IS AT ROW 2 COL 6

```
ITERATION   6
BASE VAR.    VALUE     X1      X2      X3      X4      X5      X6      X7
   X5         0.00    0.25   -8.00   -1.00    9.00    1.00    0.00    0.00
   X6         0.00    0.50  -12.00   -0.50    3.00    0.00    1.00    0.00
   X7         1.00    0.00    0.00    1.00    0.00    0.00    0.00    1.00
   -Z         0.00   -0.75   20.00   -0.50    6.00    0.00    0.00    0.00
```

FULL SIMPLEX METHOD

THERE IS NO PHASE I

```
ITERATION  |0
BASE VAR.    VALUE     X1      X2      X3      X4      X5      X6      X7
   X5         0.00    0.50  -12.00   -0.50    3.00    1.00    0.00    0.00
   X6         0.00    0.25   -8.00   -1.00    9.00    0.00    1.00    0.00
   X7         1.00    0.00    0.00    1.00    0.00    0.00    0.00    1.00
   -Z         0.00   -0.75   20.00   -0.50    6.00    0.00    0.00    0.00
```

PIVOT IS AT ROW 1 COL 1

```
ITERATION   1
BASE VAR.    VALUE     X1      X2      X3      X4      X5      X6      X7
   X1         0.00    1.00  -24.00   -1.00    6.00    2.00    0.00    0.00
   X6         0.00    0.00   -2.00   -0.75    7.50   -0.50    1.00    0.00
   X7         1.00    0.00    0.00    1.00    0.00    0.00    0.00    1.00
   -Z         0.00    0.00    2.00   -1.25   10.50    1.50    0.00    0.00
```

PIVOT IS AT ROW 3 COL 3

```
ITERATION   2
BASE VAR.    VALUE     X1      X2      X3      X4      X5      X6      X7
   X1         1.00    1.00  -24.00    0.00    6.00    2.00    0.00    1.00
   X6         0.75    0.00   -2.00    0.00    7.50   -0.50    1.00    0.75
   X3         1.00    0.00    0.00    1.00    0.00    0.00    0.00    1.00
   -Z         1.25    0.00    2.00    0.00   10.50    1.50    0.00    1.25
```

FINAL SOLUTION

MINIMUM OF Z =   -1.25

CONSTRAINT	BASIS	VALUE	STATE	SLACK
1	1	1.00	BINDING	0.00
2	6	0.75	SLACK	0.75
3	3	1.00	BINDING	0.00

Exercises 2

1 Use the Simplex Method to maximise $3x_1 + 6x_2 + 2x_3$ for $x_1, x_2, x_3 \geqslant 0$ and

$$3x_1 + 4x_2 + x_3 \leqslant 2$$
$$x_1 + 3x_2 + 2x_3 \leqslant 1.$$

2 Find non-negative x_1, x_2, x_3 such that

$$- x_2 + 4x_3 \geqslant 1$$
$$- x_1 + 5x_2 \qquad \leqslant 1$$
$$- x_1 + 2x_2 + 3x_3 \leqslant 9$$

and $\qquad 2x_1 - 2x_2 + 4x_3 = z$ is a minimum.

3 A firm manufactures three products (A, B, C), each of which requires time on all of the four manufacturing facilities I, II, III, IV. The manufacturing times and profit margins per unit amount of the products are shown below:

	\multicolumn{4}{c}{Time (hours)}	
	I	II	III	IV	Profit ($)
A	1	3	1	2	3
B	6	1	3	3	6
C	3	3	2	4	4

If the production times available on the facilities, I, II, III and IV are 84, 42, 21 and 42 hours respectively, determine which products should be made and in what quantities. (You may assume there to be an unlimited market for each product, set-up times prior to a change of product being manufactured are negligible and maximisation of profit is the only consideration.)

4 A manufacturer of soft drinks has two bottling machines A and B. Machine A is designed for $\frac{1}{2}$ litre bottles and machine B for 1 litre bottles, but each can be used for both sizes of bottle with some loss in efficiency as shown in the table, which gives the rates at which the machines work.

Machine	$\frac{1}{2}$ litre bottles	1 litre bottles
A	50 per minute	20 per minute
B	40 per minute	30 per minute

Both machines run for 6 hours each day of a 5 day week. The profit on a $\frac{1}{2}$ litre bottle is 4 cents and on a 1 litre bottle 10 cents. The weekly production cannot exceed 50 000 litres and the market will only take at most 44 000 $\frac{1}{2}$ litre bottles and 30 000 1 litre bottles.

The manufacturer wishes to use his bottling plant so as to maximise his profit. Formulate this problem as a linear programming problem and find an optimum solution.

5 Minimise $z = 50x_1 + 25x_2$ for $x_1, x_2 \geqslant 0$ which satisfy

$$x_1 + 3x_2 \geqslant 8$$
$$3x_1 + 4x_2 \geqslant 19$$
$$3x_1 + x_2 \geqslant 7.$$

6 Maximise $w = 8y_1 + 19y_2 + 7y_3$ for $y_1, y_2, y_3 \geqslant 0$ which satisfy

$$y_1 + 3y_2 + 3y_3 \leqslant 50$$
$$3y_1 + 4y_2 + y_3 \leqslant 25.$$

7 A manufacturer of central heating components makes radiators in 4 models. The constraints on his production are the limits on his labour force (in man hours), and the steel sheet from which the radiators are pressed. The sheet is delivered each week by a regular supplier. The data in the table give information on the four models.

Radiator model	A	B	C	D	Available
Man hours needed	0·5	1·5	2	1·5	500 hours
Steel sheet (m²) needed	4	2	6	8	2500 m²
Profit/radiator ($)	5	5	12·5	10	

He sets up the problem as a linear programming problem with profit maximisation as his objective. Obtain the problem he formulates and solve it using the Simplex Method.

8 A small firm produces two types of bearing, A and B, which each have to be processed on three machines, namely lathes, grinders and drill presses. The time taken for each stage in the production process is shown in the table below.

Bearing type	Time required (in hours)			Profit per bearing
	Lathe	Grinder	Drill press	
A	0·01	0·02	0·04	80c
B	0·02	0·01	0·01	125c
Total time available per week in hours	160	120	150	

The firm wishes to produce bearings in quantities in order to maximise its profit. Formulate this problem as a linear programming problem and obtain the solution using the Simplex Method.

Verify the solution graphically.

9 Minimise $z = -4x_1 - 5x_2 - 9x_3 - 11x_4$ for $x_1, x_2, x_3, x_4 \geqslant 0$ which satisfy

$$x_1 + x_2 + x_3 + x_4 \leqslant 15$$
$$7x_1 + 5x_2 + 3x_3 + 2x_4 \leqslant 120$$
$$3x_1 + 5x_2 + 10x_3 + 15x_4 \leqslant 100.$$

10 A firm can advertise its products using four media, television, radio, newspapers and posters. From various advertising experiments which they have carried out in the past they estimate that there are increased profits of $10, $3, $7 and $4 per dollar spent on advertising via these media.

The allocation of the advertising budget to the various media is subject to the following restrictions:

(i) The total budget must not exceed $500 000.
(ii) The policy is to spend at most 40% of the budget on television and at least 20% on posters.
(iii) Because of the appeal of the products to teenagers the policy is to spend at least half as much on radio as on television.

Formulate the problem of allocating the available money to the various media as a linear programming problem and use the Simplex Method to obtain a solution.

11 Find $x_1, x_2 \geqslant 0$ such that

$$2x_1 - x_2 \leqslant 4$$
$$x_1 - 2x_2 \leqslant 2$$
$$x_1 + x_2 \leqslant 5$$

and
$$-x_1 + x_2 = z \text{ is minimised.}$$

12 Oxfam tries to construct a diet for refugees containing at least 20 units of protein, 30 units of carbohydrates, 10 units of fat and 40 units of vitamins. What is the cheapest way to achieve this, given the prices and contents per kg (or per litre) of the 5 available foods are as shown below?

	Bread	Soya meat	Dried fish	Fruit	Milk (sub.)
Protein	2	12	10	1	2
Carbohydrates	12	0	0	4	3
Fat	1	8	3	0	4
Vitamins	2	2	4	6	2
Cost	12	36	32	18	10

13 The ESBP oil company is involved in purchasing crude oil from a number of different sources, W, X, Y and Z, and refining it into different grades, A, B and C, of lubricating oil ready for sale. The specifications for the grades of lubricating oil impose restrictions on the proportions of each crude oil which may be used in the refining process. There are also limitations to the amount of each grade of lubricating oil which can be sold.

	Specifications	Market available (gallons)
Grade A	At least 10% W At most 25% Z	90 000
Grade B	At least 15% W	100 000
Grade C	At least 20% X At most 50% Y	120 000

The costs of the crude oil and the selling price for each grade of lubricating oil are given in the table below.

Cost per gallon	Selling price per gallon
W 75	A 90
X 72	B 87
Y 60	C 84
Z 67	

Assuming that the crude oil is available in unlimited quantities formulate the problem of maximising profit as a linear programming problem and find the optimum solution.

14   A weaving shop has to be manned 24 hours per day by weavers according to the following table:

Time of day	2–6	6–10	10–14	14–18	18–22	22–02
Minimum number of weavers required	4	8	10	7	12	4

Each weaver is to work eight consecutive hours per day. The objective is to find the smallest number of weavers required to comply with the above requirements.

Show that the manning schedule is subject to constraints which may be written

$$
\begin{aligned}
x_1 \qquad\qquad\quad + x_6 - x_7 \qquad\qquad\qquad\qquad &= 4 \\
x_1 + x_2 \qquad\qquad\qquad - x_8 \qquad\qquad\qquad &= 8 \\
x_2 + x_3 \qquad\qquad\qquad - x_9 \qquad\qquad &= 10 \\
x_3 + x_4 \qquad\qquad\qquad - x_{10} \qquad &= 7 \\
x_4 + x_5 \qquad\qquad\qquad - x_{11} \quad &= 12 \\
x_5 + x_6 \qquad\qquad\qquad - x_{12} &= 4
\end{aligned}
$$

Explain the physical significance of each variable and show that $\{x_1, x_2, x_3, x_5, x_7, x_{12}\}$ form a feasible basis. Express this linear programming problem in canonical form and hence obtain an optimum solution using the Simplex Method.

15 Minimise $-3x_1 - 4x_2 - 5x_3$ for non-negative x_1, x_2, x_3 which satisfy

$$x_1 + \quad x_2 + \quad x_3 \geqslant 4$$
$$2x_1 + 3x_2 + 4x_3 \leqslant 6.$$

16 Modify the programs given so that the variable chosen to enter the basis at the next iteration is the *first* one found to have a negative coefficient. Try your program on the problems of these exercises. Does the change lead to any marked increase in computational time?

17 The non-negativity conditions in an L.P. program can be generalised to $l_i \leqslant x_i \leqslant u_i$ where l_i and u_i are lower and upper bounds for the variable x_i. These could clearly be included as extra constraints.

There is, however, a short cut procedure which is based on modifying the rules of the Simplex Method. This is discussed in Chapter 11 of the book by W. W. Garvin (see the references). It is also treated in the paper by G. B. Dantzig; 'Upper bounds, secondary constraints and block triangularity in Linear Programming', *Econometrica*, **23**, 174–183 (1955).

Consult these references and try to incorporate the necessary modifications in the programs. Do the short cut procedures improve on the crude idea of just treating the conditions as additional constraints?

3

Sensitivity Analysis

3.1 The Inverse of the Basis and the Simplex Multipliers

It was pointed out (refer to equation (2.7)) that the canonical form for a particular basis could be obtained by premultiplying the original constraints by the inverse of that basis. Thus for the general L.P. problem with m equation constraints in n non-negative variables, of the form

$$Ax = b, \tag{3.1}$$

if B represents the m columns of A which correspond to the basic variables, so that A can be written

$$A = (BR) \tag{3.2}$$

where B is the $m \times m$ matrix of the basis and R the $m \times (n - m)$ matrix of the non-basic variables, then the canonical form for the basis is obtained by multiplying

$$(BR)x = b \tag{3.3}$$

by B^{-1} to obtain

$$(I_m B^{-1}R)x = B^{-1}b = b' \tag{3.4}$$

which will represent the canonical form for the constraints.

Of course the Simplex Method does not *appear* to calculate the inverse matrix by direct inversion. It proceeds iteratively but it does in fact calculate the inverse of the basis and this inverse can be found by examining the tableaux of the simplex calculations.

If a_j represents the column of coefficients of the variable x_j in the first **equation form** of the constraints, then

$$a_j' = B^{-1}a_j \tag{3.5}$$

will represent the column of coefficients of x_j in the canonical form. If x_j is a slack variable that arose from a '$\leqslant$' constraint then

$$a_j = \begin{pmatrix} 0 \\ 0 \\ 1 \\ 0 \\ 0 \end{pmatrix} \leftarrow p\text{th row say}$$

so that a_j will be the pth column of B^{-1}. If x_k is a slack variable that arose from a '$\geqslant$' constraint then

$$a_k = \begin{pmatrix} 0 \\ 0 \\ -1 \\ 0 \\ 0 \end{pmatrix} \leftarrow q\text{th row say}$$

so that a_k will be the *negative* of the qth column of B^{-1}.

As an illustration consider the first and last tableaux of Example 1 of Section 2.1. These are reproduced below.

Iteration	Basis	Value	x_1	x_2	x_3	x_4	
0	x_3	1700	3	4	1	.	First tableau
	x_4	1600	2	5	.	1	
	$-z$	0	-2	-4	.	.	

2	x_1	300	1	.	$\frac{5}{7}$	$-\frac{4}{7}$	Final optimum tableau
	x_2	200	.	1	$-\frac{2}{7}$	$\frac{3}{7}$	
	$-z$	1400	.	.	$\frac{2}{7}$	$\frac{4}{7}$	

The optimal basis is (x_1, x_2). The matrix of coefficients of the basis in the first form for the constraints is

$$B = \begin{pmatrix} 3 & 4 \\ 2 & 5 \end{pmatrix}.$$

In the first canonical form the matrix of coefficients of (x_3, x_4) is

$$I_2 = \begin{pmatrix} 1 & 0 \\ 0 & 1 \end{pmatrix}.$$

In the final tableau it is

$$B^{-1}I_2 = B^{-1} = \begin{pmatrix} \frac{5}{7} & -\frac{4}{7} \\ -\frac{2}{7} & \frac{3}{7} \end{pmatrix}$$

and it is easy to check that this is indeed B^{-1}.

In the same way, consider the first and last tableaux of Example 1 of Section 2.3. The slack variables (x_3, x_4, x_5, x_6) have matrix of coefficients in the first tableau

$$\begin{pmatrix} -1 & 0 & 0 & 0 \\ 0 & -1 & 0 & 0 \\ 0 & 0 & 1 & 0 \\ 0 & 0 & 0 & 1 \end{pmatrix}.$$

The final basis (x_1, x_2, x_3, x_4) has matrix of coefficients in the first tableau

$$
B = \begin{pmatrix} 1 & 0 & -1 & 0 \\ 0 & 1 & 0 & -1 \\ 1 & 1 & 0 & 0 \\ -1 & 4 & 0 & 0 \end{pmatrix}.
$$

In the final tableau the slack variables (x_3, x_4, x_5, x_6) have matrix

$$
\begin{pmatrix} 0 & 0 & \frac{4}{5} & -\frac{1}{5} \\ 0 & 0 & \frac{1}{5} & \frac{1}{5} \\ 1 & 0 & \frac{4}{5} & -\frac{1}{5} \\ 0 & 1 & \frac{1}{5} & \frac{1}{5} \end{pmatrix}
$$

so that B^{-1} is (note the sign change in the first two columns)

$$
\begin{pmatrix} 0 & 0 & \frac{4}{5} & -\frac{1}{5} \\ 0 & 0 & \frac{1}{5} & \frac{1}{5} \\ -1 & 0 & \frac{4}{5} & -\frac{1}{5} \\ 0 & -1 & \frac{1}{5} & \frac{1}{5} \end{pmatrix}
$$

as is readily verified.

The reader is urged to look back at other examples and pick out the inverse of the basis. The values of the basic variables are of course given by $b' = B^{-1}b$ and this can also be checked.

In each canonical form the basic variables appropriate to that form have been eliminated from the objective function z. The Simplex Method does this in an iterative manner. It is possible to imagine it being done at each stage by using the **first form** for the constraints. For the general problem

$$
\left. \begin{array}{l} a_{11}x_1 + a_{12}x_2 + \ldots + a_{1n}x_n = b_1 \\ a_{21}x_1 + a_{22}x_2 + \ldots + a_{2n}x_n = b_2 \\ \cdots\cdots\cdots\cdots\cdots\cdots\cdots\cdots \\ a_{m1}x_1 + a_{m2}x_2 + \ldots + a_{mn}x_n = b_m \end{array} \right\} \tag{3.6}
$$

$$c_1x_1 + c_2x_2 + \ldots + c_nx_n = z \text{ is to be minimised.}$$

We can multiply the constraints by numbers $\pi_1, \pi_2, \ldots, \pi_m$ and add to z to obtain

$$
x_1\left(c_1 + \sum_{i=1}^{m} a_{i1}\pi_i\right) + x_2\left(c_2 + \sum_{i=1}^{m} a_{i2}\pi_i\right) + \ldots + x_n\left(c_n + \sum_{i=1}^{m} a_{in}\pi_i\right) = z + \Sigma b_i\pi_i. \tag{3.7}
$$

We can choose the π_i so that the coefficients of the basic variables in equation (3.7) are zero. The π_i are known as the **Simplex Multipliers.** If $x_1, x_2, \ldots, x_m$ are basic (there is no loss of generality here) the π_i are determined from

$$
\left. \begin{array}{l} a_{11}\pi_1 + a_{21}\pi_2 + \ldots + a_{m1}\pi_m = -c_1 \\ a_{12}\pi_1 + a_{22}\pi_2 + \ldots + a_{m2}\pi_m = -c_2 \\ \cdots\cdots\cdots\cdots\cdots\cdots\cdots\cdots \\ a_{1m}\pi_1 + a_{2m}\pi_2 + \ldots + a_{mm}\pi_m = -c_m \end{array} \right\}
$$

i.e.
$$B^T\boldsymbol{\pi} = -c_B \tag{3.8}$$

where B is the matrix of coefficients of the basic variables and $c_B^\mathrm{T} = (c_1, \ldots, c_m)$ the coefficients of the basic variables in the first form for z.

$$\boldsymbol{\pi} = \begin{pmatrix} \pi_1 \\ \pi_2 \\ \vdots \\ \pi_m \end{pmatrix}.$$

Thus

$$\boldsymbol{\pi} = -(\boldsymbol{B}^{-1})^\mathrm{T} \boldsymbol{c}_B \tag{3.9}$$

However, the values of $\pi_1, \ldots, \pi_m$ are perhaps more easily obtained by inspecting the Simplex tableaux.

Suppose x_j is a slack variable which does not appear in the first form for the objective function. If x_j arises from the pth constraint which is a '$\leqslant$' constraint, its coefficient in the pth constraint when the problem is put in standard form will be $+1$. x_j will not occur anywhere else. Thus it is clear that its coefficient in the optimal form for z will be π_p. Similarly if x_k is the slack in the qth constraint which is a '$\geqslant$' constraint, its coefficient in the optimal form for z will be $-\pi_q$.

Consider once again the tableaux of Example 1 of Section 2.1. The coefficients of x_3 and x_4 in the optimal form for z are $\frac{2}{7}$ and $\frac{4}{7}$ respectively and these are the Simplex multipliers π_1, π_2 for the optimal basis.

The first form for the constraints and objective function is

$$
\begin{aligned}
3x_1 + 4x_2 + x_3 \quad\quad &= 1700 \times \pi_1 \, (= \tfrac{2}{7}) \\
2x_1 + 5x_2 \quad\quad + x_4 &= 1600 \times \pi_2 \, (= \tfrac{4}{7}) \\
-2x_1 - 4x_2 \quad\quad\quad\ &= z.
\end{aligned}
$$

Multiply the constraints by π_1 and π_2 as shown and add to z to obtain

$$x_1(-2 + 3 \times \tfrac{2}{7} + 2 \times \tfrac{4}{7}) + x_2(-4 + 4 \times \tfrac{2}{7} + 5 \times \tfrac{4}{7}) + \tfrac{2}{7}x_3 + \tfrac{4}{7}x_4$$

$$= z + 1700 \times \tfrac{2}{7} + 1600 \times \tfrac{4}{7}$$

i.e.

$$\tfrac{2}{7}x_3 + \tfrac{4}{7}x_4 = z + 1400 \tag{3.10}$$

which is indeed the final form for z.

Now from equation (3.7) in the final form for z the coefficients of the basic variables will be zero by choice of the π_i and the coefficients of the non-basic variables will be positive. Also since the non-basic variables are zero each term on the L.H.S. of equation (3.7) is zero; either the variable or its coefficient is zero. Thus the optimal value of z is given by

$$z^{\mathrm{opt}} + \Sigma b_i \pi_i = 0$$

i.e.

$$z^{\mathrm{opt}} = -\Sigma b_i \pi_i. \tag{3.11}$$

This is evident for the example we have just considered (see equation (3.10)).

For Example 1 of Section 2.3 the coefficients of x_3, x_4, x_5, x_6 in the final tableau are $(0, 0, \frac{16}{5}, \frac{1}{5})$ and the Simplex multipliers are:

$$(-0, -0, \tfrac{16}{5}, \tfrac{1}{5})$$

(note the sign change for the first two which as it happens has no importance in this case).

Verify that multiplication of the constraints in (2.22) by $0, 0, \frac{16}{5}, \frac{1}{5}$ followed by addition to z as given in equations (2.22) does indeed generate the final form for z and that

$$-(-0 \times 10 + -0 \times 5 + \tfrac{16}{5} \times 20 + \tfrac{1}{5} \times 20) = -68.$$

The reader is urged to repeat this type of exercise with the other examples. It will improve his understanding of 'what is going on'.

We shall see in the next section that the π_i have an important economic interpretation in many problems. The Simplex multipliers and the inverse of the optimal basis play an important role in the process of investigating how the solution changes if our problem is slightly changed.

3.2 The Effect of Changes in the Problem

We have seen that L.P. problems result from many different practical situations. The examples and exercises of the previous two chapters have given some indication of the scope of the method. The coefficients which occur in the mathematical problem often have a physical significance in the practical problem. The coefficients in the objective function may represent the profit to be derived from the various 'activities'. The R.H.S. values in the constraints may represent the limits placed on the availability of limited resources. In such cases it will be appreciated that these values could change and this in turn will change the mathematical problem. For instance, by working overtime it may be possible to increase machine production hours, a fire in a warehouse may further restrict the supply of raw materials, shortages due to bad weather may increase the profit to be made on certain items. How do we cope with this situation?

One very crude way is to change the mathematical problem to take account of the physical changes and solve the new problem 'from scratch'. However, this procedure may be very inefficient and not take account of the useful work that has already been done in solving the problem before the changes.

We shall consider in turn

(i) Changes in the b_i (the R.H.S. values).

(ii) Changes in the c_j (the coefficients in the objective function).

(iii) The inclusion of more variables.

(iv) The addition of extra constraints.

(i) *Changes in the b_i*

The original constraints are assumed to be

$$Ax = b \quad \text{with} \quad z = c^T x \text{ to be minimised.}$$

Suppose the new problem is

$$Ax = b + \Delta b \quad \text{where} \quad \Delta b = \begin{pmatrix} \Delta b_i \\ \Delta b_2 \\ \vdots \\ \Delta b_m \end{pmatrix},$$

with the same objective function $z = c^T x$.

Suppose we have solved the original problem. Suppose for the optimal basis, its matrix of coefficients in A is B with inverse B^{-1}. Suppose that the Simplex multipliers are $\pi_1, \pi_2, ..., \pi_m$. The values of the basic variables in the original problem will be given by

$$x_B = B^{-1}b = b' \geqslant 0. \tag{3.12}$$

Also from equation (3.7) the value of the objective function will be given by

$$z^{\text{opt}} = - \Sigma b_i \pi_i \tag{3.13}$$

and each of

$$c_j + \sum_{i=1}^{m} a_{ij} \pi_i \geqslant 0 \tag{3.14}$$

(the coefficients of the basic variables are 0, of the non-basic variables $\geqslant 0$).

Now if only the b_i change, equation (3.14) will still be valid for the new problem. Thus provided the same basic solution is also feasible for the new problem, it will be the optimal basic feasible solution for this problem. The new values for these basic variables will simply be

$$x_B^* = B^{-1}(b + \Delta b) = b' + B^{-1}\Delta b. \tag{3.15}$$

If this is positive it will be a basic feasible solution for the new problem and also the optimal solution from equation (3.14). The new value of z will be

$$z^* = - \sum_{i=1}^{m} (b_i + \Delta b_i)\pi_i. \tag{3.16}$$

Thus from equation (3.13) we may obtain

$$\frac{\partial z^{\text{opt}}}{\partial b_i} = - \pi_i \tag{3.17}$$

where we are regarding z^{opt} as a function of $b_1, b_2, ..., b_m$. Of course if the b_i change by too large values, there will come a point at which x_B^* as given by equation (3.15) is not feasible. Then we would have to start again.

Example 1

Let us again consider Example 1 of Section 2.1. We restate the problem for convenience.

A firm produces self-assembly bookshelf kits in two models A and B. Production of kits is limited by the availability of raw material (high quality board) and machine processing time. Each unit of A requires 3 m^2 of board and each unit of B requires 4 m^2 of board. The firm can obtain up to 1700 m^2 of board each week from its suppliers. Each unit of A needs 12 minutes of machine time and each unit of B needs 30 minutes of machine time. Each week a total of 160 machine hours is available. If the profit on each A unit is \$2 and on each B unit is \$4, how many units of each model should the firm plan to produce each week?

If the plan is for x_1 A units and x_2 B units the L.P. problem is to find $x_1, x_2 \geqslant 0$ such that

$$3x_1 + 4x_2 \leqslant 1700$$
$$\tfrac{1}{5}x_1 + \tfrac{1}{2}x_2 \leqslant 160$$

i.e.

$$3x_1 + 4x_2 \leqslant 1700$$
$$2x_1 + 5x_2 \leqslant 1600$$

which minimise $-2x_1 - 4x_2 = z$ (the negative of the profit).

The first and last tableaux are

Iteration	Basis	Value	x_1	x_2	x_3	x_4	
	x_3	1700	3	4	1	.	First tableau
	x_4	1600	2	5	.	1	
	$-z$	0	-2	-4	.	.	

	Basis	Value	x_1	x_2	x_3	x_4	
	x_1	300	1	.	$\tfrac{5}{7}$	$-\tfrac{4}{7}$	Optimal tableau
	x_2	200	.	1	$-\tfrac{2}{7}$	$\tfrac{3}{7}$	
	$-z$	1400	.	.	$\tfrac{2}{7}$	$\tfrac{4}{7}$	

The inverse of the basis is

$$\begin{pmatrix} \tfrac{5}{7} & -\tfrac{4}{7} \\ -\tfrac{2}{7} & \tfrac{3}{7} \end{pmatrix}.$$

The simplex multipliers are $\tfrac{2}{7}, \tfrac{4}{7}$.

A   Suppose we can buy extra board from a second timber merchant. How much per square metre are we prepared to pay for it?

Thus we suppose that the 1700 in the first constraint changes to 1701. The new vector of b's will be

$$\begin{pmatrix} 1701 \\ 1600 \end{pmatrix}.$$

The new values of the basic variables will be

$$\begin{pmatrix} x_1^* \\ x_2^* \end{pmatrix} = \begin{pmatrix} \tfrac{5}{7} & -\tfrac{4}{7} \\ -\tfrac{2}{7} & \tfrac{3}{7} \end{pmatrix} \begin{pmatrix} 1701 \\ 1600 \end{pmatrix} = \begin{pmatrix} 300 \\ 200 \end{pmatrix} + \begin{pmatrix} \tfrac{5}{7} & -\tfrac{4}{7} \\ -\tfrac{2}{7} & \tfrac{3}{7} \end{pmatrix} \begin{pmatrix} 1 \\ 0 \end{pmatrix}$$

$$= \begin{pmatrix} 300 + \tfrac{5}{7} \\ 200 - \tfrac{2}{7} \end{pmatrix} \text{ and this is feasible.}$$

The optimal value for z changes to $(-\Sigma b_i \pi_i)$

$$-(\tfrac{2}{7} \times 1701 + \tfrac{4}{7} \times 1600) = -1400 - \tfrac{2}{7}.$$

Thus the **profit** will increase by $\$\tfrac{2}{7}$ and this is the maximum price we would be prepared to pay for the extra square metre of board. Otherwise we should lose on the deal. This value is of course π_1.

B   Suppose we can obtain extra machine time by working overtime. If this costs $\$7$ per hour extra is it worth it?

In this case the vector $\mathbf{b}$ changes to

$$\begin{pmatrix} 1700 \\ 1610 \end{pmatrix}$$

in the mathematical problem. The new values of the basic variables are

$$\begin{pmatrix} x_1^+ \\ x_2^+ \end{pmatrix} = \begin{pmatrix} \tfrac{5}{7} & -\tfrac{4}{7} \\ -\tfrac{2}{7} & \tfrac{3}{7} \end{pmatrix} \begin{pmatrix} 1700 \\ 1610 \end{pmatrix} = \begin{pmatrix} 300 - \tfrac{40}{7} \\ 200 + \tfrac{30}{7} \end{pmatrix} \text{ which is feasible.}$$

The optimal value for z changes to

$$-(\tfrac{2}{7} \times 1700 + \tfrac{4}{7} \times 1610) = -1400 - \tfrac{40}{7}.$$

The profit will increase by $\$\tfrac{40}{7}$. Since the hour's overtime costs $\$7$ it is not worth it.

It is easy to see that solving the new problem from scratch will reproduce the results we have obtained. But it is not necessary to start from scratch. Indeed in large problems it would be very inefficient.

Note that in A the new value for z is

$$-2(300 + \tfrac{5}{7}) - 4(200 - \tfrac{2}{7})$$

and in B it is

$$-2(300 - \tfrac{40}{7}) - 4(200 + \tfrac{30}{7}).$$

(ii) *Changes in the c_j*

Example 2

Suppose in Example 1 that the profit on each A unit is $\$P_1$ and on each B unit is $\$P_2$. For what possible values of P_1 and P_2 is the solution we have obtained optimal?

The objective function in the first tableau is given by $-P_1 x_1 - P_2 x_2 = z + 0$. Since only the last line has been changed the canonical form for the constraints of the basis will be as given, i.e.,

$$x_1 \qquad + \tfrac{5}{7} x_3 - \tfrac{4}{7} x_4 = 300$$
$$x_2 - \tfrac{2}{7} x_3 + \tfrac{3}{7} x_4 = 200.$$

To obtain z in canonical form we have to eliminate x_1 and x_2 from z.

We can do this by multiplying the first constraint (in the final tableau) by P_1, the second by P_2 and adding to z to obtain

$$(\tfrac{5}{7} P_1 - \tfrac{2}{7} P_2)x_3 + (-\tfrac{4}{7} P_1 + \tfrac{3}{7} P_2)x_4 = z + 300 P_1 + 200 P_2.$$

Thus the solution will be optimal provided the coefficients of the non-basic variables are both positive. Thus the solution is optimal provided

$$\tfrac{5}{7} P_1 - \tfrac{2}{7} P_2 \geqslant 0 \quad \text{and} \quad -\tfrac{4}{7} P_1 + \tfrac{3}{7} P_2 \geqslant 0$$

$$\therefore \quad \frac{P_1}{P_2} \geqslant \tfrac{2}{5} \quad \text{and} \quad \frac{P_1}{P_2} \leqslant \tfrac{3}{4}.$$

This is clear from Fig. 1.1 where B will be optimal provided the contour of z through B lies between the two constraint lines which intersect at B.

This approach is very useful in analysing the effect of changes in the c_j on the solution of the problem. The values of the basic variables and the canonical form for the constraints are unchanged. If the coefficients of the non-basic variables in the new objective function are positive the solution is optimal. If one or more of them are negative we shall need to bring this variable into the basis. We shall then have a suitable canonical form for the whole problem to continue with further iterations of the Simplex Method. The earlier calculations will not be wasted.

(iii) The Inclusion of Additional Variables

Example 3

Suppose a third type of kit (C), can be made. Suppose it uses $4 \, \text{m}^2$ of board and needs 20 minutes of machine time. If the profit on each unit is $\$P$, should we make it?

If we make x_5 units of C our problem in standard form becomes: find x_1, x_2, x_3, x_4, x_5 such that

$$3x_1 + 4x_2 + x_3 + + 4x_5 = 1700$$
$$\tfrac{1}{5} x_1 + \tfrac{1}{2} x_2 + x_4 + \tfrac{1}{3} x_5 = 160$$
$$- 2x_1 - 4x_2 - Px_5 = z \text{ is a minimum}$$

i.e.
$$3x_1 + 4x_2 + x_3 + 4x_5 = 1700$$
$$2x_1 + 5x_2 + x_4 + \tfrac{10}{3} x_5 = 1600$$
$$- 2x_1 - 4x_2 - Px_5 = z.$$

In the final tableau the first two rows of the x_5 column will be, by equation (3.5)

$$B^{-1} \begin{pmatrix} 4 \\ \tfrac{10}{3} \end{pmatrix} = \begin{pmatrix} \tfrac{5}{7} & -\tfrac{4}{7} \\ -\tfrac{2}{7} & \tfrac{3}{7} \end{pmatrix} \begin{pmatrix} 4 \\ \tfrac{10}{3} \end{pmatrix} = \begin{pmatrix} \tfrac{20}{21} \\ \tfrac{6}{21} \end{pmatrix}.$$

Since the Simplex multipliers are

$$\begin{pmatrix} \pi_1 \\ \pi_2 \end{pmatrix} = \begin{pmatrix} \tfrac{2}{7} \\ \tfrac{4}{7} \end{pmatrix}$$

then by equation (3.7) the coefficient of x_5 in the canonical form for z will be

$$- P + \tfrac{2}{7} \times 4 + \tfrac{4}{7} \times \tfrac{10}{3} = - P + \tfrac{64}{21}.$$

The final tableau will be (the only change is the x_5 column)

Iteration	Basis	Value	x_1	x_2	x_3	x_4	x_5
	x_1	300	1	.	$\frac{5}{7}$	$-\frac{4}{7}$	$\frac{20}{21}$
	x_2	200	.	1	$-\frac{2}{7}$	$\frac{3}{7}$	$\frac{6}{21}$
	$-z$	1400	.	.	$\frac{2}{7}$	$\frac{4}{7}$	$-P+\frac{64}{21}$

If $-P+\frac{64}{21} \geqslant 0$ then this solution is optimal and x_5 remains non-basic and we do not make the new model. If, however,

$$-P+\frac{64}{21} < 0, \quad \text{i.e.} \quad P > \frac{64}{21}$$

then we should make x_5 basic. We 'pick up' the computation from the canonical form just given and continue with the Simplex Method.

(iv) *The Addition of Extra Constraints*

Example 4

Suppose that during a period of economic recession the sales team report that the market will not stand more than 550 sales of kits each week. How does this affect the production schedule?

This constraint on total sales is equivalent to

$$x_1 + x_2 \leqslant 550.$$

This additional constraint has to be included in the problem. However, in this particular instance it has no effect on the optimal solution. In this solution $x_1 = 300$ and $x_2 = 200$ and $x_1 + x_2 = 500$ and so satisfies the extra constraint.

Had the recession been more severe with total sales restricted to at most 450 weekly, things would be different. We take up this point in the next section.

3.3 The Dual Simplex Method

Example 1

Suppose, following on from the last section that total weekly sales are restricted to at most 450 kits.

Thus the additional constraint

$$x_1 + x_2 \leqslant 450$$

has to be included. In equation form this can be written as

$$x_1 + x_2 \quad + x_5 = 450$$

where the slack variable $x_5 \geqslant 0$.

This constraint is violated by the optimal solution to the original problem. Are we forced to solve the problem from scratch with this constraint included at the outset? Had we done this and repeated the previous calculations the additional constraint would have finished up expressed in terms of the non-basic variables. We can deduce the form for this from the current canonical form,

$$x_1 \quad + \tfrac{5}{7} x_3 - \tfrac{4}{7} x_4 \quad = 300$$
$$x_2 - \tfrac{2}{7} x_3 + \tfrac{3}{7} x_4 \quad = 200.$$

Thus
$$x_1 + x_2 \quad\quad\quad + x_5 = 450$$

takes the form (on eliminating x_1 and x_2)

$$-\tfrac{3}{7} x_3 + \tfrac{1}{7} x_4 + x_5 = -50.$$

Then our last tableau would have been (the only new feature is the form of the additional constraint):

Iteration	Basis	Value	x_1	x_2	x_3	x_4	x_5
2	x_1	300	1	.	$\tfrac{5}{7}$	$-\tfrac{4}{7}$	.
	x_2	200	.	1	$-\tfrac{2}{7}$	$\tfrac{3}{7}$	.
	x_5	-50	.	.	$-\tfrac{3}{7}$	$\tfrac{1}{7}$	1
	$-z$	1400	.	.	$\tfrac{2}{7}$	$\tfrac{4}{7}$	.

At this point we encounter a problem. In this canonical form for the basis x_1, x_2, x_5, the form for the objective is optimal since all the coefficients are non-negative, but the basis is *not* feasible. x_5 is negative. Is there a way despite this to salvage the useful work that has been done so far? There is, and the procedure is that of the **Dual Simplex Method.**

The Simplex Method can be regarded as a procedure which starts with positive values for the basic variables, and whilst retaining this feature transforms to a canonical form (in stages perhaps) in which all the coefficients in the objective function eventually become positive. The Dual Simplex Method works in reverse. It does not insist at the outset that the basic variables are positive but does require that (for a minimisation problem) the coefficients in the objective function are all non-negative. Whilst retaining this latter feature it transforms the constraints to obtain an all positive basis eventually, at which point the minimum will have been obtained.

In our problem the basic variable x_5 is negative and is a candidate to be removed *from* the basis. What variable is to replace it? In the x_5 row we seek a negative pivot and one such that in the subsequent transformation (which again will take the form of equations (2.15), ..., (2.20)) the coefficients in the objective function will remain positive. Before formalising these rules let us see how it works out for our problem. In the x_5 row there is only one negative coefficient, the coefficient of x_3 which is $-\tfrac{3}{7}$. If we divide the equation by $-\tfrac{3}{7}$ to make x_3 the basic variable (with coefficient 1) we obtain

$$x_3 - \tfrac{1}{3} x_4 - \tfrac{7}{3} x_5 = \tfrac{350}{3}.$$

This certainly makes the value of x_3 positive. The next step is to eliminate x_3 from the other constraints and the objective. These are just Simplex type computations and can be carried out in tableau form as shown. The pivot is the negative value $-\frac{3}{7}$ which is asterisked.

Iteration	Basis	Value	x_1	x_2	x_3	x_4	x_5
2	x_1	300	1	.	$\frac{5}{7}$	$-\frac{4}{7}$	.
	x_2	200	.	1	$-\frac{2}{7}$	$\frac{3}{7}$	.
	x_5	-50	.	.	$-\frac{3}{7}*$	$\frac{1}{7}$	1
	$-z$	1400	.	.	$\frac{2}{7}$	$\frac{4}{7}$	.
3	x_1	$\frac{650}{3}$	1	.	.	$-\frac{1}{3}$	$\frac{5}{3}$
	x_2	$\frac{700}{3}$	.	1	.	$\frac{1}{3}$	$-\frac{2}{3}$
	x_3	$\frac{350}{3}$	.	.	1	$-\frac{1}{3}$	$-\frac{7}{3}$
	$-z$	$\frac{4100}{3}$	.	.	.	$\frac{2}{3}$	$\frac{2}{3}$

The final tableau gives the optimal solution for the new problem. This give

$$x_1 = \tfrac{650}{3}, \quad x_2 = \tfrac{700}{3} \quad \text{with} \quad z = -\tfrac{4100}{3}.$$

Since in this solution x_3 is basic there is slack in the supply of raw material and the knock-on effect might be that we reduce our order from the supplier of board.

The procedure we have described can be generalised. If the objective function has all positive coefficients the steps are as follows.

1   Find a negative basic variable. If there is none we have the optimal solution; if there is more than one find the most negative. Suppose this variable is the basic variable in the rth constraint. This gives the variable to come out of the basis.

2   In row r look for *negative* coefficients a'_{rj}. If there are none there is no feasible solution to the problem. (This is clear if you think about it carefully.) For negative coefficients a'_{rj} in this row find the

$$\min_{j} \left| \frac{c'_j}{a'_{rj}} \right|.$$

If this arises in column s, variable s is the variable to enter the basis.

3   Carry out the usual Simplex transformation with a'_{rs} as pivot. From equation (2.19) in the next tableau

$$c^{+}_j = c'_j - \frac{c'_s a'_{rj}}{a'_{rs}}$$

and since a'_{rs} is negative, and all c'_j are positive, this is positive, since s arose from

$$\min_{j} \left| \frac{c'_j}{a'_{rj}} \right| = \left| \frac{c'_s}{a'_{rs}} \right|.$$

Thus the Dual Simplex Method differs from the Simplex Method only in the way in which it selects the variables to leave and enter (in that order) the basis.

The reader is asked to check Example 1 geometrically. He will then realise that with the Dual Simplex Method we may start with a point *outside* the feasible region. The method eventually generates a feasible point which is also optimal.

The Dual Simplex procedure has been incorporated into the following program which can be used either to carry out the Simplex Method if the basis is feasible, or the Dual Simplex Method if the coefficients in the objective function are all positive. The program is virtually identical with the first program given apart from the routines 1 and 2 just discussed in connection with the Dual Simplex Method. It assumes that we have a canonical form but one in which basic variables may be negative. If it is used to incorporate an extra constraint into a previously solved problem the extra constraint must be put in canonical form first.

```
PROGRAM SimplexandDualSimplex (input,output);
CONST
   m=3; n=5;        { No. of constraints and total no. of variables }   {**}
   fwt=8;  dpt=2;  { Output format constants for tableau values }       {**}
   fwi=1;           { Output format constant for indices }              {**}
   largevalue = 1.0E20;  smallvalue = 1.0E-10;                          {**}

TYPE   mrange = 1..m;  nrange = 1..n;
   matrix = ARRAY [mrange,nrange] OF real;
   column = ARRAY [mrange] OF REAL;
   baseindex = ARRAY [mrange] OF integer;
   row = ARRAY [nrange] OF real;
   rowboolean = ARRAY [nrange] OF boolean;

VAR
   a : matrix;  { Matrix A in standard form of problem, see (2.3) }
   b : column;  { Vector b in standard form of problem, see (2.3) }
   c : row;     { Coefficients of objective function, see (2.1) }
   basic : baseindex;        { Basic variables at each stage }
   nonbasic : rowboolean;  { Status indicators for variables }
   z0 : real;                { Value of objective function }
   it : integer;             { Iteration counter }
   solution, unbounded,      { Iteration process terminators }
   nonfeasible : boolean;
   r, s : integer;           { Row and column of pivot element }

PROCEDURE inputdata;
VAR  i, j : integer;
BEGIN
  FOR i:=1 TO m DO
  BEGIN FOR j:=1 TO n DO read(a[i,j]);  read(b[i]) END;
  FOR j:=1 TO n DO read(c[j]);  read(z0);
  FOR i:=1 TO m DO read(basic[i]); { Initial base }
END; { inputdata }

PROCEDURE initialise;
VAR  i, j : integer;
BEGIN  it:=0;  unbounded:=false;  nonfeasible:=false;
  FOR j:=1 TO n DO nonbasic[j]:=true;
  FOR i:=1 TO m DO nonbasic[basic[i]]:=false
END; { initialise }
```

```
PROCEDURE outputtableau;
VAR  i, j : integer;
BEGIN  writeln; writeln('     ITERATION ', it:2);
  write('      BASE VAR.   ', ' ':fwt-5, 'VALUE');
  FOR j:=1 TO n DO write(' ':fwt-fwi-1, 'X', j:fwi); writeln;
  FOR i:=1 TO m DO
  BEGIN
    write(' ':8-fwi, 'X', basic[i]:fwi, ' ':7, b[i]:fwt:dpt);
    FOR j:=1 TO n DO write(a[i,j]:fwt:dpt); writeln
  END;
  write(' ':7, '-Z', ' ':7, z0:fwt:dpt);
  FOR j:=1 TO n DO write(c[j]:fwt:dpt);  writeln
END; { outputtableau }




PROCEDURE nextbasicvariable (VAR r,s:integer);
VAR  i, j : integer;  min : real;
BEGIN  min:=largevalue;  { Decide whether to employ Simplex }
  FOR j:=1 TO n DO        { Method or Dual Simplex Method.   }
    IF nonbasic[j] THEN IF c[j]<min THEN BEGIN min:=c[j]; s:=j END;
  solution := c[s] > -smallvalue;
  IF NOT solution THEN  { Find pivot in column s using Simplex Method }
  BEGIN  unbounded:=true; i:=1;   { Check that at least one value }
    WHILE unbounded AND (i<=m) DO { in column s is positive.      }
    BEGIN  unbounded := a[i,s] < smallvalue;  i:=i+1 END;
    IF NOT unbounded THEN     { Variable s enters the basis, now }
    BEGIN  min:=largevalue;  { find the variable, basic[r], to  }
      FOR i:=1 TO m DO        { leave the basis.                 }
        IF a[i,s] > smallvalue THEN
          IF b[i]/a[i,s] < min THEN BEGIN min:=b[i]/a[i,s]; r:=i END
    END
  END
  ELSE  { Find pivot using Dual Simplex Method }
  BEGIN  min:=largevalue;  { Find the variable, basic[r], }
    FOR i:=1 TO m DO       { to leave the basis.          }
      IF b[i]<=min THEN BEGIN min:=b[i]; r:=i END;
    solution := b[r] > -smallvalue;
    IF NOT solution THEN
    BEGIN  min:=-largevalue;  { Find the variable, s, }
      FOR j:=1 TO n DO        { to enter the basis.   }
        IF nonbasic[j] THEN
          IF a[r,j] < -smallvalue THEN
            IF c[j]/a[r,j] > min THEN
            BEGIN  min := c[j]/a[r,j];  s:=j  END;
      nonfeasible := min < -largevalue + smallvalue
    END
  END;
  IF NOT (solution OR unbounded OR nonfeasible) THEN
  BEGIN nonbasic[basic[r]]:=true; nonbasic[s]:=false; basic[r]:=s;
    writeln; writeln('     PIVOT IS AT ROW ', r:fwi, ' COL ', s:fwi)
  END
END; { nextbasicvariable }
```

```
PROCEDURE transformtableau (r,s:integer);
{ Construct the new canonical form, implementing (2.15) to (2.20) }
VAR  i, j : integer;  pivot, savec : real;  savecol : column;
BEGIN
  FOR i:=1 TO m DO savecol[i]:=a[i,s];  savec:=c[s];  pivot:=a[r,s];
  b[r]:=b[r]/pivot;  {(2.15)}
  FOR j:=1 TO n DO a[r,j]:=a[r,j]/pivot;  {(2.16)}
  FOR i:=1 TO m DO
    IF i<>r THEN
    BEGIN b[i] := b[i] - savecol[i]*b[r];  {(2.17)}
      FOR j:=1 TO n DO a[i,j] := a[i,j] - savecol[i]*a[r,j]  {(2.18)}
    END;
  FOR j:=1 TO n DO c[j] := c[j] - savec*a[r,j];  {(2.19)}
  z0 := z0 - savec*b[r];  {(2.20)}  it := it+1
END; { transformtableau }

BEGIN  { Main Program }
  writeln; writeln('    SIMPLEX / DUAL SIMPLEX METHOD'); writeln;
  inputdata;  initialise;
  REPEAT
    outputtableau;
    nextbasicvariable(r,s);
    IF NOT (solution OR unbounded OR nonfeasible) THEN
      transformtableau(r,s)
  UNTIL solution OR unbounded OR nonfeasible;
  { Output results }  writeln;
  IF unbounded THEN writeln ('    VARIABLE ', s:fwi, ' IS UNBOUNDED')
  ELSE IF nonfeasible THEN writeln('    NO FEASIBLE SOLUTION')
      ELSE writeln('    MINIMUM AT Z=', -z0:fwt:dpt)
END. { Simplex }
```

The Dual Simplex Method is useful in allowing us to add extra constraints to a problem that has already been solved. It is valuable as a method in its own right and sometimes allows us to avoid the use of artificial variables.

Example 2

Find non-negative x_1, x_2 such that

$$x_1 + 2x_2 \geqslant 6$$
$$2x_1 + x_2 \geqslant 6$$
$$7x_1 + 8x_2 \leqslant 56$$

for which $x_1 + x_2 = z$ is a minimum.

In standard form the problem becomes, find $x_i \geqslant 0$ such that

$$x_1 + 2x_2 - x_3 \qquad\qquad = 6$$
$$2x_1 + x_2 \quad - x_4 \qquad = 6$$
$$7x_1 + 8x_2 \qquad\quad + x_5 = 56$$

for which $x_1 + x_2 \qquad\qquad = z$ is a minimum.

With basis x_3, x_4, and x_5 (i.e. x_1, x_2 non-basic) the objective function is in optimal form. Of course the basis is not feasible since $x_3 = -6$ and $x_4 = -6$. If we multiply these two constraints by -1 (just to get the correct form for the basis) we have

$$- x_1 - 2x_2 + x_3 \qquad\qquad = -6$$
$$- 2x_1 - x_2 \quad\;\; + x_4 \qquad = -6$$
$$7x_1 + 8x_2 \qquad\quad + x_5 = 56$$
$$x_1 + x_2 \qquad\qquad\quad = z.$$

We have avoided the use of artificial variables in the first two constraints. We can proceed directly using the Dual Simplex Method, the successive tableau being as shown.

Iteration	Basis	Value	x_1	x_2	x_3	x_4	x_5
0	x_3	-6	-1	-2	1	.	.
	x_4	-6	-2^*	-1	.	1	.
	x_5	56	7	8	.	.	1
	$-z$	0	1	1	.	.	.
1	x_3	-3	.	$-\frac{3}{2}^*$	1	$-\frac{1}{2}$	.
	x_1	3	1	$\frac{1}{2}$	.	$-\frac{1}{2}$	.
	x_5	35	.	$\frac{9}{2}$	.	$\frac{7}{2}$	1
	$-z$	-3	.	$\frac{1}{2}$	.	$\frac{1}{2}$	.
2	x_2	2	.	1	$-\frac{2}{3}$	$\frac{1}{3}$	.
	x_1	2	1	.	$\frac{1}{3}$	$-\frac{2}{3}$	.
	x_5	26	.	.	3	2	1
	$-z$	-4	.	.	$\frac{1}{3}$	$\frac{1}{3}$	.

This is the optimal solution. The optimal form for z has been preserved throughout and now the basis is feasible. Thus the minimum for z is 4 when $x_1 = 2$ and $x_2 = 2$. This can be verified graphically.

The values in the CONST part of the PROGRAM *SimplexandDualSimplex* are appropriate for this problem and the data file would contain the following values:

$$-1 \quad -2 \quad 1 \quad 0 \quad 0 \quad -6 \quad -2 \quad -1 \quad 0 \quad 1 \quad 0 \quad -6$$
$$7 \quad 8 \quad 0 \quad 0 \quad 1 \quad 56 \quad 1 \quad 1 \quad 0 \quad 0 \quad 0 \quad 0 \quad 3 \quad 4 \quad 5$$

The computer output which clearly reproduces the tableaux above follows.

```
SIMPLEX / DUAL SIMPLEX METHOD

ITERATION  0
BASE VAR.      VALUE       X1        X2        X3        X4        X5
   X3          -6.00     -1.00     -2.00      1.00      0.00      0.00
   X4          -6.00     -2.00     -1.00      0.00      1.00      0.00
   X5          56.00      7.00      8.00      0.00      0.00      1.00
   -Z           0.00      1.00      1.00      0.00      0.00      0.00

PIVOT IS AT ROW 2 COL 1

ITERATION  1
BASE VAR.      VALUE       X1        X2        X3        X4        X5
   X3          -3.00      0.00     -1.50      1.00     -0.50      0.00
   X1           3.00      1.00      0.50      0.00     -0.50      0.00
   X5          35.00      0.00      4.50      0.00      3.50      1.00
   -Z          -3.00      0.00      0.50      0.00      0.50      0.00

PIVOT IS AT ROW 1 COL 2

ITERATION  2
BASE VAR.      VALUE       X1        X2        X3        X4        X5
   X2           2.00      0.00      1.00     -0.67      0.33      0.00
   X1           2.00      1.00      0.00      0.33     -0.67      0.00
   X5          26.00      0.00      0.00      3.00      2.00      1.00
   -Z          -4.00      0.00      0.00      0.33      0.33      0.00

MINIMUM AT Z=    4.00
```

Exercises 3

1 A manufacturer produces two products, P which sells at $2000 per tonne, and Q which sells at $1000 per tonne. These can be manufactured from either of two raw materials, A which costs $600 per tonne, and B which costs $900 per tonne. Each 100 tonnes of A yields 30 tonnes of P and 50 tonnes of Q, whilst each 100 tonnes of B yields 60 tonnes of P and 10 tonnes of Q. If the manufacturer processes x tonnes of A and y tonnes of B, show that his profit is

$$\$(500x + 400y).$$

The factory can handle not more than 10 000 tonnes of raw material each year. The suppliers of raw material can supply not more than 6000 tonnes of A and not more than 8000 tonnes of B per year. The manufacturer can sell up to 5000 tonnes of P and up to 3200 tonnes of Q each year. Find what quantities of A and B should be ordered to maximise his profit, and show that his profit is then $4 550 000.

The suppliers of A are threatening a price increase. By how much would the price have to rise before the manufacturer would be justified in changing his order?

2 Consider the problem; find $x_i \geqslant 0$ such that

$$-x_1 + 10x_2 + x_3 \qquad = 40$$
$$x_1 + \ x_2 \qquad + x_4 = 20$$

which minimise $f = 10x_1 - 111x_2$.

Find the optimal simplex tableau.

The extra constraint

$$x_3 + x_4 \geqslant 5$$

is found to be necessary.

Use the Dual Simplex Method to solve the enlarged problem starting from the optimal solution of the original problem.

3 Consider the problem

$$\text{Min } z = \sum_{j=1}^{n} c_j x_j \quad \text{for} \quad x_j \geqslant 0 \text{ such that}$$

$$\sum_{j=1}^{n} a_{ij} x_j \geqslant b_i \ (i = 1, ..., m).$$

Show that (after having introduced slack variables into the constraints) the simplex multipliers π_i associated with the optimal solution, must satisfy

$$c_j + \sum_{i=1}^{m} a_{ij} \pi_i \geqslant 0 \quad j = 1, ..., n \tag{A}$$

$$-\pi_i \geqslant 0 \quad i = 1, ..., m \tag{B}$$

[N.B. $-\pi_i$ is the coefficient of x_{n+i} in the optimal form for z.]

Verify that (B) is correct from the result $\partial(z \text{ opt})/\partial b = -\pi_i$.

4 A factory can produce three basic commodities. One article of type I requires 3 units of raw material A and 1 unit of raw material B and gives a profit of 3 units. One article of type II requires 4 units of raw material A and 3 units of raw material B and gives a profit of 6 units. One article of type III requires 1 unit of material A, 2 units of B and gives a profit of 2 units. If 20 units of A and 10 units of B are all that are available find the optimal production schedule. If one more unit of material A (or B) became available what is the highest price we would be prepared to pay for it?

5 The profit per unit of A, B, C is 5, 3, 4 units respectively. One unit requires a certain time on each of the machines I and II as indicated in the table. If 12 hours of production time and 15 hours of production time are the daily availability of the machines I and II, find the optimum production schedule. Impute the value of an extra hour's production time on I (II).

Time on I, II required for 1 unit of A, B, C		
A	B	C
I	3	2	3
II	4	1	2

6   A firm of cabinet makers specialises in the production of sideboards. It can manufacture three types of sideboard that require different amounts of labour at each of the three stages of production. This is shown below.

	Labour (man hours) required to produce one sideboard of type		
	A	B	C
Sawmill	1	2	4
Assembly shop	2	4	2
Finishing shop	1	1	2

The available labour each week for the sawmill is 360 man hours, for the assembly shop 520 man hours and for the finishing shop 220 man hours. The profit on each sideboard of type A, B, C is $9, $11, $15 respectively. Show that the problem of obtaining the optimum production schedule can be put in the form: find $x_i \geqslant 0$ such that

$$x_1 + 2x_2 + 4x_3 + x_4 \qquad\qquad = 360$$
$$2x_1 + 4x_2 + 2x_3 \qquad + x_5 \qquad = 520$$
$$x_1 + x_2 + 2x_3 \qquad\qquad + x_6 = 220$$

and $\qquad w = 9x_1 + 11x_2 + 15x_3 \qquad\qquad$ is a maximum.

Explain the physical significance of each of the x_i and use the Simplex Method to show that in the optimal solution

$$x_1 = 180, \quad x_2 = 40, \quad x_3 = 0.$$

Impute the value of an extra man hour of labour in the sawmill, the assembly shop, the finishing shop, on the basis of this solution.

   The production manager refuses to accept the above solution and points out that in order to meet its obligations to the highly valued hotel trade the company must produce at least 10 sideboards of type C each week. How does this extra requirement influence the solution?

7   A university can accommodate not more than 5000 students in its lecture rooms and laboratories. It is prevented by the government from recruiting more than 4000 home students, but it can take as many overseas students as it wishes.

   The university has 440 academic staff. It allows one member of staff for every twelve home students, and one for every ten overseas students. It guarantees that at least 40% of home students and 80% of overseas students will have places in halls of residence, where there are 2800 places available. The university receives £2000 per annum from government sources for each home student and charges a fee of £3000 per annum to overseas students.

   Assuming that the university's sole aim is to maximise its income, find how many home and overseas students it should aim to recruit. Show that the maximum income it can receive is £11 850 000 per annum.

The university could employ additional staff at a cost of £10 000 per annum each. Would this be worthwhile?

8 A firm can manufacture four products at its factory. Production is limited by the machine hours available and by the number of special components available. The necessary data are given below.

	Product				
	1	2	3	4	Availability
Machine hours/unit	1	3	8	4	90 hours/day
Components/unit	2	2	1	3	80 components/day
Production cost ($/unit)	20	25	40	55	
Sales income ($/unit)	30	45	80	85	

The daily production x_j of product j is given by the solution of the problem

minimise
$$z = -x_1 - 2x_2 - 4x_3 - 3x_4$$

subject to
$$x_1 + 3x_2 + 8x_3 + 4x_4 \leqslant 90$$
$$2x_1 + 2x_2 + x_3 + 3x_4 \leqslant 80$$
$$x_j \geqslant 0, \quad j = 1, ..., 4.$$

Explain the derivation of this problem.

The optimal tableau as given by a computer print-out is given below.

Basis	Value	x_1	x_2	x_3	x_4	x_5	x_6
x_1	10	1	$-\frac{1}{5}$	-4	0	$-\frac{3}{5}$	$\frac{4}{5}$
x_4	20	0	$\frac{4}{5}$	3	1	$\frac{2}{5}$	$-\frac{1}{5}$
$-z$	70	0	$\frac{1}{5}$	1	0	$\frac{3}{5}$	$\frac{1}{5}$

(i) What are the simplex multipliers?

(ii) What is the inverse of the basis?

(iii) The firm can increase the available machine hours by 10 hours/day by hiring extra machinery. The cost of this would be $40 per day. Should they rent it and if so what is the new production schedule?

(iv) The price of one of the raw materials used in the manufacture of products 1 and 3 is highly variable. At present its price is $80 per 10 kg. Product 1 uses $1\frac{1}{4}$ kg per unit and product 3 uses $2\frac{1}{2}$ kg per unit. The price of this material has been included in the production cost above. Within what range of values can the price of this material fluctuate if the original solution is to remain valid?

(v) An industrial dispute at the works of one of the consumers of product 4 means that daily production of this product must be restricted to 15 units. Use the Dual Simplex Method to find the new production schedule.

9   Find $x_1, x_2 \geqslant 0$ which satisfy

$$x_1 + 3x_2 \geqslant 8$$
$$3x_1 + 4x_2 \geqslant 19$$
$$3x_1 + x_2 \geqslant 7$$

and minimise $50x_1 + 25x_2$.
Do not use artificial variables.

10   Find $x_1, x_2 \geqslant 0$ which satisfy

$$7x_1 + 8x_2 \geqslant 56$$
$$x_1 + 2x_2 \leqslant 6$$

and minimise $x_1 + x_2 = z$.

11   Show directly that for the general L.P. problem if changes Δb_i in the b's cause changes Δx_j in the x's then

$$\Delta z = -\sum_{i=1}^{m} \pi_i \Delta b_i = \sum_{j=1}^{n} c_j \Delta x_j$$

provided the basis remains feasible.

12   In seeking to optimise an industrial process excess temperature (x_1), and feed rate (x_2) are thought to be the important variables which determine the output. After some discussion with the process engineer you as operational research consultant formulate the problem in linear programming terms as:
Find x_1, x_2, both non-negative such that

$$x_1 + x_2 \leqslant 6$$
$$4x_1 + 11x_2 \leqslant 44$$

which also minimise $z = -2x_1 - 5x_2$. Use the Simplex Method to obtain the solution to this problem.

On being asked to implement this solution the plant foreman expresses considerable doubts. After further discussions and to the considerable embarrassment of the process engineer it is decided that the constraint

$$x_1 + 2x_2 \leqslant 8$$

should be included.

Show by carrying out the necessary calculations how the Dual Simplex procedure allows the final tableau of the earlier calculation to be used in obtaining the solution to the new problem.

13   Use the program for the Simplex/Dual Simplex Method to write a computer program which will solve an L.P. problem with GC '$\geqslant$' constraints and LC '$\leqslant$' constraints. The program should read the constraints and the objective function coefficients and set up a first basic solution (not necessarily feasible since we do not want to use artificial variables). It should then use the Simplex Method and/or the Dual Simplex Method to solve the problem.

14 Use your program to solve the problem:
find $x_1, x_2, x_3 \geqslant 0$ such that

$$3x_1 - 2x_2 + 4x_3 \geqslant 8$$
$$x_1 + 2x_2 + x_3 \leqslant 9$$
$$2x_1 + x_2 \leqslant 6$$

and
$$3x_1 + x_2 + 2x_3 = z \text{ is minimised.}$$

15 Suppose that in solving an L.P. problem using the Simplex Method, the canonical form at iteration k is given by equations (2.4) and (2.5) in matrix form. Suppose x_s enters the basis and the basic variable x_r in the rth constraint leaves the basis. Show that premultiplication of the canonical form by the matrix

rth column

$$F = r\text{th row} \quad \begin{pmatrix} 1 & 0 & 0 & -\dfrac{a'_{1s}}{a'_{rs}} & 0 & 0 & 0 \\[2mm] 0 & 1 & 0 & -\dfrac{a'_{2s}}{a'_{rs}} & & & \\[2mm] 0 & 0 & & \dfrac{1}{a'_{rs}} & 0 & 0 & 0 \\[2mm] 0 & 0 & & -\dfrac{a'_{ms}}{a'_{rs}} & 0 & 1 & 0 \\[2mm] 0 & 0 & & -\dfrac{c'_{s}}{a'_{rs}} & 0 & 0 & 1 \end{pmatrix} \left.\vphantom{\begin{pmatrix} 1 \\ 1 \\ 1 \\ 1 \\ 1 \end{pmatrix}}\right\} m+1 \text{ rows}$$

$\underbrace{}$
$m + 1$ columns

generates the next canonical form.

If B_k^{-1} is the inverse of the basis at the kth iteration and B_{k+1}^{-1} is the inverse of the basis at the $(k+1)$th iteration show that

$$B_{k+1}^{-1} = EB_k^{-1}$$

and write down the matrix E.

4

The Transportation Problem

4.1 The Nature of the Problem and its Solution

Example 1

A company which runs a chain of Department Stores wishes to transport some beds from its 3 warehouses to 5 of its retail outlets. There are 15, 25, 20 beds respectively at the warehouses and the 5 stores need 20, 12, 5, 8 and 15 beds respectively. The costs ($'s) of moving 1 bed from warehouse to store are given in the table.

		To				
		S_1	S_2	S_3	S_4	S_5
From	W_1	1	0	3	4	2
	W_2	5	1	2	3	3
	W_3	4	8	1	4	3

How should the distribution be planned so as to minimise the costs? Let x_{ij} be the number of beds sent from warehouse i to store j. Clearly $x_{ij} \geqslant 0$ and because of the limitations on supply from the warehouses and demand at the stores they satisfy:

(for supply)

$$x_{11} + x_{12} + x_{13} + x_{14} + x_{15} = 15$$
$$x_{21} + x_{22} + x_{23} + x_{24} + x_{25} = 25$$
$$x_{31} + x_{32} + x_{33} + x_{34} + x_{35} = 20$$

(for demand)

$$x_{11} \qquad + x_{21} \qquad + x_{31} \qquad = 20$$
$$x_{12} \qquad + x_{22} \qquad + x_{32} \qquad = 12$$
$$x_{13} \qquad + x_{23} \qquad + x_{33} \qquad = 5$$
$$x_{14} \qquad + x_{24} \qquad + x_{34} \qquad = 8$$
$$x_{15} \qquad + x_{25} \qquad + x_{35} = 15.$$

Subject to these constraints the cost

$$C = x_{11} + 0x_{12} + 3x_{13} + 4x_{14} + 2x_{15} + 5x_{21} + \cdots 4x_{34} + 3x_{35}$$

is to be minimised.

The problem is thus a linear programming problem, but of a special type. In particular, the coefficients in the constraints are either 0 or 1, and each variable appears in just two constraints. At first sight it might be thought that the equality signs in the constraints should be replaced by '$\leqslant$' signs in the supply constraints, and '$\geqslant$' signs in the demand constraints. However, because the total demand is equal to the total supply we must have equality in all cases. We also notice that the sum of the first three constraints yields the same result as the sum of the last five constraints. Thus we only have *seven* (not eight) independent constraints and consequently a basic feasible solution, and hence the optimal solution will contain in general seven non-zero values for the x_{ij}.

These results will generalise to the transportation problem with m supply points (supplying amounts $a_i, i = 1, 2, ..., m$) and n demand points (requiring amounts $b_j, j = 1, ..., n$) where

$$\sum_{i=1}^{m} a_i = \sum_{j=1}^{n} b_j. \tag{4.1}$$

If c_{ij} is the cost of transportation of 1 unit from supply point i to demand point j the problem will be to find $x_{ij} \geqslant 0$ which satisfy

$$
\begin{array}{llll}
x_{11} + x_{12} + \cdots + x_{1n} & & & = a_1 \\
& x_{21} + \cdots + x_{2n} & & = a_2 \\
\multicolumn{4}{c}{\cdots\cdots\cdots\cdots\cdots\cdots\cdots\cdots\cdots\cdots\cdots\cdots} \\
& & x_{m1} + \cdots + x_{mn} = a_m & \tag{4.2} \\
x_{11} & + x_{21} & + x_{m1} & = b_1 \\
\quad x_{12} & + x_{22} & + x_{m2} & = b_2 \\
\multicolumn{4}{c}{\cdots\cdots\cdots\cdots\cdots\cdots\cdots\cdots\cdots\cdots\cdots\cdots} \\
\quad x_{1n} & + x_{2n} & + x_{mn} = b_n &
\end{array}
$$

which minimise

$$C = c_{11}x_{11} + c_{12}x_{12} + \cdots \qquad + c_{mn}x_{mn}.$$

More briefly we can write equations (4.2) as, find $x_{ij} \geqslant 0$ for which

$$\sum_{j=1}^{n} x_{ij} = a_i > 0 \quad (i = 1, ..., m) \tag{4.3}$$

$$\sum_{i=1}^{m} x_{ij} = b_j > 0 \quad (j = 1, ..., n) \tag{4.4}$$

which minimise

$$C = \sum_{i=1}^{m} \sum_{j=1}^{n} c_{ij}x_{ij}. \tag{4.5}$$

Since

$$\sum_{i=1}^{m} a_i = \sum_{i=1}^{m} \sum_{j=1}^{n} x_{ij} = \sum_{j=1}^{n} \sum_{i=1}^{m} x_{ij} = \sum_{j=1}^{n} b_j$$

by equation (4.1), there are only $m + n - 1$ independent constraints and hence $m + n - 1$ basic variables in a basic feasible solution.

Rather than consider the constraints as set out, it is simpler to consider the transportation **array** in the form given below. We have to place non-negative variables in the cells in such a way that the row and column totals are as indicated, and such that the sum of the products of these variables with the costs (shown in the bottom right hand corner of each cell) is a minimum. The array shown is appropriate to example 1. It is clear how it should be generalised.

1	0	3	4	2	15
5	1	2	3	3	25
4	8	1	4	3	20
20	12	5	8	15	

In this form it is easy to see how we can construct a first basic feasible solution to the problem. We can do this by the 'lowest costs first' rule. Since we are trying to minimise the total cost we scan the cells for the lowest cost; 0 in row 1 column 2, and give x_{12} the value 12 (the lower of the row and column totals involved). This exhausts column 2 so we delete it and reduce the row total in row 1 by 12 from 15 to 3. We then apply the same process to the reduced array.

1	12 0	3	4	2	~~15~~ 3
5	1	2	3	3	25
4	8	1	4	3	20
20	~~12~~	5	8	15	

We next give x_{11} the value 3 (we could also have chosen to give x_{33} the value 5), delete row 1, reduce the column total in column 1 to 17 and proceed to the next array etc.

With a little practice it is possible to carry out this process mentally (for a reasonable sized problem). When the final variable is assigned its value, both row and column totals in which it is located will be exhausted. In this way we obtain

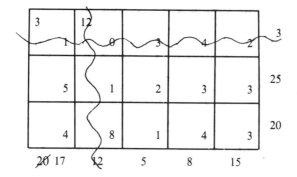

the solution (the values of the variables are in the top left hand corners of the cells), with 7 basic variables as given below. Other variables are of course zero. For the general array with m rows and n columns we shall in general have $m + n - 1$ basic variables because of equation (4.1).

3 ⌜1	12 ⌜0	⌜3	⌜4	⌜2	15
2 ⌜5	⌜1	⌜2	8 ⌜3	15 ⌜3	25
15 ⌜4	⌜8	5 ⌜1	⌜4	⌜3	20
20	12	5	8	15	

The total cost associated with this solution is

$$C = 3 \times 1 + 12 \times 0 + 2 \times 5 + 8 \times 3 + 15 \times 3 + 15 \times 4 + 5 \times 1 = \$147.$$

We now have to try to improve on this solution with a view to reducing C. Before doing this, however, we note that the specific results we have obtained here and the manner in which they have been obtained can be extended to the general transportation problem (4.2).

We find c_{sr} the smallest of the c_{ij}. Put x_{sr} equal to the smaller of a_s and b_r. If this is a_s, delete row s, replace b_r by $b_r - a_s$ and repeat the procedure on the reduced array. The basic solution obtained in this way will contain $m + n - 1$ basic variables and any one of these x_{ij} will be given by an expression of the type

$$x_{ij} = \pm \sum_{\text{some rows } p} a_p \mp \sum_{\text{some cols } q} b_q.$$

We can prove this result to be true for *all* basic feasible solutions. We first prove that all bases for the transportation problem are **triangular**. We explain this term.

A system of equations is said to be triangular if there exists at least one equation which contains just one unknown, and when this is evaluated there will again exist at least one equation with just one unknown etc.

e.g.

$$3x_1 + 2x_2 + 7x_3 + \ x_4 = 9$$
$$x_2 \qquad + 4x_4 = 2$$
$$3x_3 + \ x_4 = 7$$
$$x_4 = 2.$$

Such a system of equations can be solved by back substitution. The 'triangle of coefficients' need not all be non-zero, but the diagonal coefficients must be non-zero.

I   *All Bases for the Transportation Problem are Triangular*

If we consider the cells of our transportation array we have to show that at least one row or at least one column has only one basic variable in it, and that when this row or column is deleted the reduced array will have the same property.

First each row and each column must contain at least one basic variable. Otherwise we could not satisfy its non-zero row or column total.

For an $m \times n$ array,

if each row has at least two basic variables,

no. of basic variables $\geqslant 2m$.

If each column has at least two basic variables,

no. of basic variables $\geqslant 2n$.

Thus no. of basic variables $\geqslant m + n$.

But this is impossible since there are $m + n - 1$ basic variables.

Thus at least one row or column has only one basic variable.

If we delete that row or column and consider the reduced array we can repeat the argument so that this property is also true for the reduced array. This establishes the result that all bases are triangular.

II   *The Values of the Basic Variables are Given by Expressions of the Form:*

$$x_{ij} = \pm \sum_{\text{some } p} a_p \mp \sum_{\text{some } q} b_q. \tag{4.6}$$

Since the basis is triangular, at least one row or column contains just one variable. Thus

$$x_{pq} = a_p \quad (\text{row } p),$$

or

$$x_{pq} = b_q \quad (\text{column } q).$$

If say the former (i.e. $a_p < b_q$) we delete row p and replace b_q by $b_q - a_p$. We now repeat the argument on the reduced array.

$$x_{p'q'} = a_{p'} \quad \text{or} \quad b_{q'} \quad \text{or} \quad b_q - a_p.$$

If the procedure is repeated we find that all values of the basic variables are either of the form

$$\sum_{\text{some } p} a_p - \sum_{\text{some } q} b_q \quad \text{or} \quad - \sum_{\text{some } p} a_p + \sum_{\text{some } q} b_q.$$

As a corollary we observe that if all a_i and b_j are integers, the values of the basic variables in a basic feasible solution are also integers. Hence the optimal solution, being a basic feasible solution, since this is an L.P. problem, will have integer values.

This is a very important point. We would not have been happy with a solution to Example 1 which told us to ship $7\frac{3}{4}$ beds from ... etc. This result assures us that we cannot have this situation.

4.2   The 'Stepping Stones' Algorithm

We could solve the transportation problem by the Simplex Method directly. Such an approach would be inefficient and would fail to take account of the special structure of the constraints. We shall use an algorithm first developed by F. L. Hitchcock and sometimes referred to as the 'Stepping Stones' algorithm to solve the problem. We shall develop the solution procedure with special reference to Example 1, but we shall point out the generality of the argument, which involves the use of the Simplex Multipliers which were defined in Chapter 3.

For Example 1 we already have a basic feasible solution which was obtained by the lowest costs first procedure. We have seen that this is a general procedure which could be applied to the equations (4.2).

Suppose for this problem we do have a basic feasible solution in which some of the x_{ij} are non-zero and others are non-basic and hence zero. If $-u_i$ and $-v_j$ are the Simplex multipliers for the ith row constraint and jth column constraint appropriate to this basis, then on multiplication of the ith row by $-u_i$ and the jth column by $-v_j$ and addition to C we obtain

$$
\begin{aligned}
x_{11} + x_{12} + \cdots + x_{1n} &= a_1(\times - u_1)\\
x_{21} + \cdots + x_{2n} &= a_2(\times - u_2)\\
&\;\vdots\\
x_{m1} + \cdots + x_{mn} &= a_m(\times - u_m)\\
x_{11} \qquad\quad + x_{21} \qquad\quad + x_{m1} &= b_1(\times - v_1)\\
x_{12} \qquad\quad + x_{22} \qquad\quad + x_{m2} &= b_2(\times - v_2)\\
&\;\vdots\\
x_{1n} \qquad\quad + x_{2n} \qquad\quad + x_{mn} &= b_n(\times - v_n)\\
c_{11}x_{11} + c_{12}x_{12} \qquad\qquad + c_{mn}x_{mn} &= C
\end{aligned}
$$

$$\therefore \sum_{i=1}^{m} \sum_{j=1}^{n} (c_{ij} - u_i - v_j)x_{ij} = C - \sum_{i=1}^{m} u_i a_i - \sum_{j=1}^{n} v_j b_j. \qquad (4.7)$$

Equation (4.7) is just the special form of equation (3.7) for the transportation problem. The coefficient of x_{ij} is simply $c_{ij} - u_i - v_j$ and this simple form results from the occurrence of x_{ij} in just two constraints; the ith row constraint and the jth column constraint.

Now if equation (4.7) is the canonical form for the objective function appropriate to the basis, the coefficients of the basic variables will be zero.

Thus for the occupied cells in our array

$$c_{ij} - u_i - v_j = 0 \qquad (4.8)$$

and these equations allow us to determine u_i and v_j.

There are $m\,u_i$'s and $n\,v_j$'s and since there are $m + n - 1$ occupied cells (basic variables) equation (4.8) will give $m + n - 1$ equations for $m + n$ unknowns. We can get a solution by giving one of them an arbitrary value (say 0) and solving for the others. This will always be possible. In Example 1 we have the following basic feasible solution at the first step.

$3 + w$ 1	$12 - w$ 0	3	4	2	15 (−4)
$2 - w$ 5	w −3 1	0 2	8 3	15 3	25 (0) $C = 147$
15 4	8	5 1	4	3	20 (−1)
20 (5)	12 (4)	5 (2)	8 (3)	15 (3)	

There are 8 unknowns $u_1, u_2, u_3,$ and $v_1, \ldots, v_5$ and 7 occupied cells. If we put $u_2 = 0$ we obtain from the 3 occupied cells in this row $v_1 = 5$, $v_4 = 3$ and $v_5 = 3$. From $v_1 = 5$ we obtain from the occupied cells in this column, $u_1 = -4$ and $u_3 = -1$. Finally from $u_1 = -4$ we obtain $v_2 = 4$ and from $u_3 = -1$ we obtain $v_3 = 2$. Thus we obtain the values shown in brackets for the u_i and v_j.

We can now test if the solution is optimal. If c'_{ij} is the coefficient of x_{ij} in the canonical form for the objective function then from equation (4.7)

$$c'_{ij} = c_{ij} - u_i - v_j. \qquad (4.9)$$

$c'_{ij} = 0$ for the basic variables. For the non-basic variables a negative value for c'_{ij} indicates that x_{ij} can be brought into the basis and bring about a decrease in the objective function. We thus calculate c'_{ij} for the unoccupied cells and note those cells in which it is negative. This is shown in the bottom left hand corner in the array. A zero value at this stage would indicate variables which will leave the objective unchanged if they enter the basis.

Thus we see that for our problem increasing x_{22} will decrease the objective. Indeed each unit increase in x_{22} will decrease the objective by 3. If we increase x_{22} to w we must decrease x_{21} by w to maintain the row (2) sum. In order to maintain the column (1) sum we must increase x_{11} by w and then to maintain the row (1) sum we must decrease x_{12} by w and then we have come full circle and everything balances. [Note we do not follow a route: put $x_{22} = w$, decrease x_{42} by w to $8 - w$, since there is no way to maintain the column (4) sum without introducing another variable. Then we would not have a basic solution. We must avoid such 'blind alleys' in the program.]

Thus we see that the maximum value we can give to w is 2. This will cause x_{21} to become non-basic and zero in the next basic feasible solution which is shown in the next array.

5 + w (1)	10 − w (0)	(3)	(4)	[0] (2)	15 (−1)
(5)	2 + w (1)	(2)	8 (3)	15 − w (3)	25 (0)
15 − w (4)	(8)	5 (1)	[−1] (4)	w [−2] (3)	20 (2)
20 (2)	12 (1)	5 (−1)	8 (3)	15 (3)	

For this array

$$C = 5 \times 1 + 10 \times 0 + 2 \times 1 + 8 \times 3 + 15 \times 3 + 15 \times 4 + 5 \times 1 = 141$$
$$= 147 - 3 \times 2$$
$$= \text{Previous value} + c'_{22} \times w.$$

We next determine the u_i and v_j for this solution. The reader should verify that these bracketed values are as shown. The c'_{ij} for the non-basic variables when 0 or negative have also been put in the bottom L.H. corner of a cell for the array. These indicate that x_{35} should enter the basis. If this is increased to w the other basic variables must be adjusted as shown. The reader should also trace this through. This indicates that the maximum value for w is 10 which will cause x_{12} to become non-basic in the next array.

15 (1)	(0)	(3)	(4)	(2)	15 (−3)
(5)	12 (1)	(2)	8 (3)	5 (3)	25 (0)
5 (4)	(8)	5 (1)	(4)	10 (3)	20 (0)
20 (4)	12 (1)	5 (1)	8 (3)	15 (3)	

C is now reduced to

$$C = 15 \times 1 + 12 \times 1 + 8 \times 3 + 5 \times 3 + 5 \times 4 + 5 \times 1 + 10 \times 3 = 121$$
$$= 141 - 2 \times 10$$
$$= \text{Previous value} + c'_{35} \times w.$$

For this array the u_i and v_j are now calculated. All the c'_{ij} for the unoccupied cells are positive. Thus this is the optimal solution in which

$$x_{11} = 15, \quad x_{22} = 12, \quad x_{24} = 8, \quad x_{25} = 5, \quad x_{31} = 5, \quad x_{33} = 5, \quad x_{35} = 10.$$

The minimum value for C is \$121.

Incidentally it might be noted that at each stage the value of C is given by

$$C = \sum_{i=1}^{m} a_i u_i + \sum_{j=1}^{n} v_j b_j. \tag{4.10}$$

This follows directly from equation (4.7), for on the L.H.S. of this equation each term is zero; either the variable is basic in which case its coefficient is zero or else the variable is non-basic and hence zero. Of course equation (4.10) is just the special case of equation (3.11) for the transportation problem.

The reader should verify the truth of equation (4.10) for each of the arrays. It can be a useful check on the calculations at each stage in the iterative process.

It is also worth remarking that for this problem x_{12} is zero in the optimal solution even though the cost for this cell is zero; hardly the result one would expect but it is certainly so.

4.3   Unbalance and Degeneracy in the Transportation Problem

The condition (4.1) played an important role in the transportation problem. For an $m \times n$ array it means that there are $m + n - 1$ basic variables in a basic feasible solution. Suppose this balance between supply and demand does not hold.

Example 1

(Adapted from Example 1 of Sections 4.1 and 4.2.)

Suppose the 15, 25 and 20 beds at the warehouses W_1, W_2, W_3 are to be used to supply 4 stores whose requirements are for 20, 12, 5 and 9 beds. Suppose the cost of moving 1 bed from warehouse to store is as given

		To			
		S_1	S_2	S_3	S_4
	W_1	2	2	2	4
From	W_2	3	1	1	3
	W_3	3	6	3	4

(The costs have been changed from the first example.)

How should the distribution be planned to minimise the cost?

The warehouses can supply 60 beds. The stores only require 46 beds. The 'trick' here is to introduce a dummy store which requires 14 beds. The transportation costs to this are put at zero. In the final solution if any beds are to be transported to this dummy store we shall ignore them. Those beds will remain in the warehouse. In this way we create a transportation problem for which equation (4.1) is true.

The first basic feasible solution for this array and the subsequent arrays leading to the final solution are as shown. The reader should check through the various calculations that have been carried out at each iteration. In fact only two iterations are required.

15					15 (-1)
2	2	2	4	0	
$5 - w$	12	5	$3 + w$		25 (0) $C = 95$
3	1	1	3	0	
w -1			$6 - w$	14	20 (1)
3	6	3	4	0	
20 (3)	12 (1)	5 (1)	9 (3)	14 (-1)	

x_{31} enters basis. Max. $w = 5$. x_{21} leaves basis.

15					15 (0)
2	2	2	4	0	
	12	5	8		25 (0) $C = 90$
3	1	1	3	0	
5			1	14	20 (1)
3	6	3	4	0	
20 (2)	12 (1)	5 (1)	9 (3)	14 (-1)	

This gives the optimum. The 14 beds in cell $(3, 5)$ remain in warehouse 3. The stores are all fully supplied.

$$x_{11} = 15, \quad x_{22} = 12, \quad x_{23} = 5, \quad x_{24} = 8, \quad x_{31} = 5, \quad x_{34} = 1.$$
$$C = \$90.$$

It is clear how we could cope with unbalance if demand exceeded supply. We would introduce a dummy supplier with zero costs. Quantities 'supplied' by this supplier in the final solution would *not* in fact be supplied. The demand they represent would not be satisfied.

Degeneracy will arise in a transportation problem if one or more of the basic variables becomes zero. From the result (4.6) we see that a degenerate solution might arise if partial sums of the row totals are equal to partial sums of the column totals. The problem can be overcome by 'almost' ignoring it. We must be very careful at all stages to distinguish between basic variables which are zero, and constitute occupied cells, and non-basic variables.

The problem might arise in the construction of our first basic feasible solution when both a row and a column total which are equal are reduced to zero. In this case we must delete only *one* of them from further consideration. The other will be deleted subsequently by assigning a value of 0 to a basic variable. In this way, since at each step, apart from the last, we delete *just one* row or column, we will obtain $m + n - 1$ basic variables and hence occupied cells as required, even if some of the basic variables are zero.

The problem might arise when we are improving upon a basic feasible solution. The rules might reduce more than one of our basic variables to zero. In this case it is important to remember that only one of them will become non-basic. The others must be retained in the basis with value zero. They constitute occupied cells for the purpose of determining the u_i's and v_j's.

Example 2

A government department has received the following tenders from three firms F_1, F_2 and F_3 for three sizes S_1, S_2 and S_3 of service overcoat.

		Price per coat in dollars Size of coat		
		S_1	S_2	S_3
	F_1	110	115	126
Firm	F_2	107	115	130
	F_3	104	109	116

Contracts have to be awarded for 1000 coats of size S_1, 1500 of size S_2 and 1200 of size S_3, but the limited production capacities of the firms mean that the total orders placed must not exceed 1000 coats with F_1, 1500 with F_2 and 2500 with F_3. Government policy stipulates that the contracts must be let so as to minimise the total cost, but that subject to this proviso, the orders should be spread as evenly as possible amongst the firms. How should the orders be placed to meet these requirements?

For this problem we notice that the total supply capacity of 5000 coats exceeds the total demand of 3700 coats. Thus we introduce a 'fictitious' category coat with demand 1300 in order to make our problem a transportation problem. The cost associated with this category will be zero and any demand for this category which appears in the final solution will be ignored. It will simply represent surplus supply capacity.

It is convenient to work in units of 100 coats.

Thus our array for the problem becomes:

	S_1	S_2	S_3	S_4	
F_1	110	115	126	0	10
F_2	107	115	130	0	15
F_3	104	109	116	0	25
	10	15	12	13	

A first basic feasible solution, the u_i, v_j, etc., and the first iteration of the calculation are shown below.

S_1	S_2	S_3	S_4	
110	115	10 126 -2	0	10 (-4)
-3 107	13 + w 115	2 - w 130 -6	0	15 (0) $C = 4273$ scaled units
10 104	2 - w 109	w -8 116	13 0	25 (-6)
10 (110)	15 (115)	12 (130)	13 (6)	

The maximum value for w is 2 and this will cause both x_{32} and x_{23} (in an obvious notation) to become zero. Only one, say x_{23}, is removed from the basis. The next basic feasible solution with its u_i and v_j appears below. It contains 6 basic variables although one of them is zero.

S_1	S_2	S_3	S_4	
-4 110	-4 115	10 - w 126	w -10 0	10 (10)
-3 107	15 115	130 -6	0	15 (6) $C = 4257$ scaled units
10 104	0 109	2 + w 116	13 - w 0	25 (0)
10 (104)	15 (109)	12 (116)	13 (0)	

The maximum value for w is 10.

110	115	126	10 0	10 (0)
−3 107	15 − w 115	130	w −6 0	15 (6) $C = 4157$
10 104	0 + w 109	12 116	3 − w 0	25 (0)
10 (104)	15 (109)	12 (116)	13 (0)	

Max. $w = 3$ and our next solution is not degenerate.

0 110	0 115	126	10 0	10 (6)
w −3 107	12 − w 115	130	3 0	15 (6) $C = 4139$
10 − w 104	3 + w 109	12 116	0	25 (0)
10 (104)	15 (109)	12 (116)	13 (−6)	

Max. $w = 10$.

110	w 0 115	126	10 − w 0	10 (0)
10 107	2 − w 115	130	3 + w 0	15 (0) $C = 4109$
104	13 109	12 116	0	25 (−6)
10 (107)	15 (115)	12 (122)	13 (0)	

This is optimal. However, since c'_{12} is zero this variable could also be brought into the basis without changing C. The maximum for w is 2.

	2		8	10 (6)
110	115	126	0	
10			5	15 (6) $C = 4109$
107	115	130	0	
	13	12		25 (0)
104	109	116	0	
10 (101)	15 (109)	12 (116)	13 (−6)	

Thus there are two optimal solutions each with a total cost of $410 900.

In the first F_2 supplies 1000 coats of S_1 and 200 coats of S_2
F_3 supplies 1300 coats of S_2 and 1200 coats of S_3

In the second F_1 supplies 200 coats of S_2
F_2 supplies 1000 coats of S_1
F_3 supplies 1300 coats of S_2 and 1200 coats of S_3.

The second solution appears to be the most suitable although in neither case are the orders very evenly spread.

We can avoid degeneracy by perturbing the row and column totals so that partial sums of row totals do *not* equal partial sums of column totals. In this case we might make the row totals $10 \cdot 01$, $15 \cdot 01$, $25 \cdot 01$ and the column totals 10, 15, 12 and $13 \cdot 03$. In the final solution we ignore the $0 \cdot 01$'s. The reader should carry through the calculation. It will be clear that it is virtually equivalent to what has been done.

We might note that the algorithm does not involve any division at any stage. Thus it is clear that if the a_i and b_j are integers and our first basic feasible solution is integral, so are all subsequent solutions, and hence of course the optimal solution. This integer property of all basic feasible solutions and hence the optimal solution was noted at the end of Section 4.1.

4.4   Implementing the Transportation Algorithm on the Computer

The program that follows is not trivial and the reader is urged to study it carefully. We try to break it down into its components which are then studied in more detail. T is the $m \times n$ table of cells where each cell contains fields x, c and *basic* to represent respectively the value of a basic variable, the associated cost and the status of the variable. Arrays a and b hold the supply and demand values, whilst u and v hold the computed Simplex multipliers for the row and column constraints. The steps of the last sections are first expressed in flow chart form.

Flow Chart for Transportation Algorithm

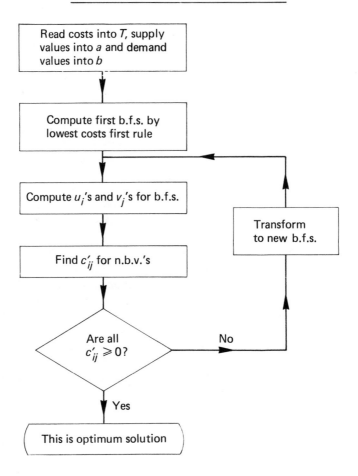

We next give a flow chart for the PROCEDURE *firstbfs* which computes the first b.f.s. This takes local copies of *a* and *b* so that the original arrays are unaltered.

Flow Chart for First b.f.s.

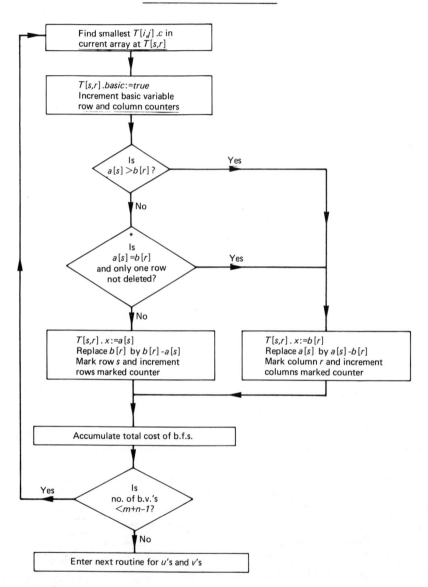

* This ensures, in the event of ties, that we do not run out of rows whilst several columns remain.

Flow Chart for the u's and v's and c_{ij}^l

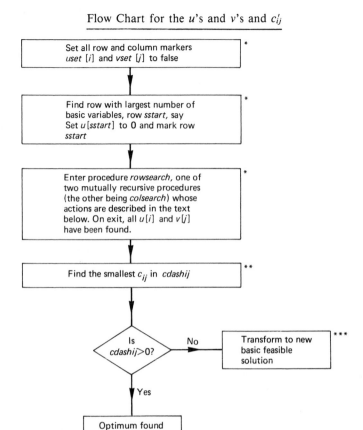

* Implemented in procedure *findusandvs*
** Implemented in main body of procedure *newbfs*
*** Implemented in procedure *modifytable*

The action of the above flow chart is implemented in the PROCEDURE *newbfs*. *rowsearch* and *colsearch* are two mutually recursive PROCEDUREs, i.e. each can call the other, which incorporate a depth-first search and backtrack technique to compute u's and v's. In *rowsearch*, when a basic variable (in row s, column j) is found and $v[j]$ has not yet been computed, the value of $v[j]$ is obtained using equation (4.8) (note that $u[s]$ is already known). *colsearch* is then called to search for basic variables in column j. *colsearch* works in a similar manner to *rowsearch*. The process finds all basic variables and hence computes all $u[i]$ and $v[j]$.

The PROCEDURE *modifytable* deserves some explanation. It finds w, the value of the newly determined basic variable, modifies other basic variables as appropriate to conserve row and column totals and finds which variable leaves the basis. The majority of this is achieved using another pair of mutually recursive PROCEDUREs, *rowpath* and *colpath*, which together search for a path through the basic variables, starting with the new basic variable and alternately examining rows and columns until the column of the new basic variable is encountered. At this point both w and the path through existing basic variables from the new basic variable have been found. Backtracking through the levels of recursion enables us to update the basic variables on the path appropriately. In *rowpath* basic variable values are reduced by w whilst in *colpath* they are increased by w. The first basic variable encountered in *rowpath* which reduces to zero is the one chosen to leave the basis. Since we have at least one basic variable in each row and column, this method guarantees to find an appropriate path. Any 'blind alley' that we get into is automatically handled by the backtracking feature of the method. The boolean array *inpath* is introduced to mark basic variables which are already in the path and thus avoid looping.

The PROCEDURE *newbfs* is completed by updating the total cost and the numbers of basic variables in rows and columns.

The Main Program calls PROCEDUREs appropriately and loops through successive basic feasible solutions until the optimum is found. A full printout can be obtained, or intermediate output suppressed depending on the first input value (positive for full printing, non-positive for suppression). The second input value determines whether the output values appear with fractional parts (input value positive) or as whole numbers (input value non-positive).

The values in the CONST part of the program listed below are appropriate for Example 1 of Section 4.1 and the data file to give the output shown would consist of the following values:

$$1 \quad -1 \quad 1 \quad 0 \quad 3 \quad 4 \quad 2 \quad 5 \quad 1 \quad 2 \quad 3 \quad 3 \quad 4 \quad 8 \quad 1 \quad 4 \quad 3$$
$$15 \quad 25 \quad 20 \quad 20 \quad 12 \quad 5 \quad 8 \quad 15.$$

The reader should follow the steps in the solution, noting that it reproduces exactly the hand worked solution of Section 4.2.

```
PROGRAM SteppingStones (input,output);
CONST
  m=3; n=5;        { No. of supply and demand constraints }        {**}
  fwt=6; dpt=1;    { Output format constants for transportation }  {**}
  fwi=1;           { array values and indices.              }      {**}
  largevalue = 1.0E20;  smallvalue = 1.0E-10;                      {**}

TYPE    mrange = 1..m;   nrange = 1..n;
  cell = RECORD  { To represent details of a single variable xij }
           x, c : real;      { Value and cost of xij }
           basic : boolean   { basic/non-basic status of xij }
         END;
  table = ARRAY [mrange,nrange] OF cell;
  col = ARRAY [mrange] OF real;   boolcol = ARRAY [mrange] OF boolean;
  row = ARRAY [nrange] OF real;   boolrow = ARRAY [nrange] OF boolean;
  rowcounter = ARRAY [mrange] OF integer;
  colcounter = ARRAY [nrange] OF integer;
```

```
VAR
  T : table;          { Main body of transportation array }
  a : col;  b : row;  { Supply and demand values, see (4.2) }
  it : integer;       { Iteration counter }
  totalcost : real;   { Total cost of current feasible solution }
  printon, fractions, { Print suppresssion and output format indicators }
  solution : boolean; { Iteration process terminator }
  basrowno : rowcounter;  { No. of basic variables in each row }
  bascolno : colcounter;  { No. of basic variables in each column }


PROCEDURE inputdata;
VAR  i, j, kp, kf : integer;
BEGIN
  read(kp);  printon := kp>0;    { Set print suppression and }
  read(kf);  fractions := kf>0;  { output format indicators  }
  FOR i:=1 TO m DO FOR j:=1 TO n DO read(T[i,j].c);  { Input costs }
  FOR i:=1 TO m DO read(a[i]);  { Input supply values }
  FOR j:=1 TO n DO read(b[j]);  { Input demand values }
END; { inputdata }


PROCEDURE initialise;
VAR  i, j : integer;
BEGIN  totalcost:=0.0;  it:=0;
  FOR i:=1 TO m DO
    FOR j:=1 TO n DO
      BEGIN  T[i,j].x:=0.0;  T[i,j].basic := false  END;
    FOR i:=1 TO m DO basrowno[i]:=0;  FOR j:=1 TO n DO bascolno[j]:=0
END; { initialise }


PROCEDURE printval (x:real);  { To output the value of x    }
BEGIN                         { in the appropriaste format }
  IF fractions THEN write(x:fwt:dpt) ELSE write(round(x):fwt)
END; { printval }


PROCEDURE outputtable;  { Output details of the first }
VAR  i, j : integer;    { basic feasible solution.    }
BEGIN  writeln('    INITIAL TABLE');  write(' ':fwi+6, 'J', '   ');
  FOR j:=1 TO n DO write(j:fwt);  writeln(' ':fwt, 'A[I]');
  writeln(' ':fwi+4, 'I', '      ');
  FOR i:=1 TO m DO
  BEGIN  write(i:fwi+5, '    ');
    FOR j:=1 TO n DO printval(T[i,j].c);  write('    ');
    printval(a[i]);  writeln
  END;  writeln;  write(' ':fwi+3, 'B[J]   ');
  FOR j:=1 TO n DO printval(b[j]);  writeln
END; { outputtable }
```

```
PROCEDURE outputbfs;   { Output details of a basic feasible }
VAR i, j : integer;   { solution, other than the first.   }
BEGIN writeln; writeln('    BASIC FEASIBLE SOLUTION NUMBER ', it:fwi);
   writeln('    BASIC VARIABLES', ' ':fwi+1, 'I', ' ':fwi+1, 'J',
            ' ':fwt, 'X[I,J]', ' ':fwt, 'C[I,J]', ' ':fwt+2, 'COST');
   FOR i:=1 TO m DO
     FOR j:=1 TO n DO WITH T[i,j] DO
       IF basic THEN
       BEGIN  write (' ':19, i:fwi+2, j:fwi+2);
          write(' ':6); printval(x); write(' ':6); printval(c);
          write(' ':6); printval(x*c); writeln
       END;
    write(' ':29+2*(fwi+fwt), 'TOTAL COST ='); printval(totalcost); writeln
END; { outputbfs }


PROCEDURE firstbfs (a:col; b:row);
{ To compute the first basic feasible solution using the 'lowest costs }
{ first' rule. Local copies of a and b will be taken, thus preserving  }
{ the original values for later use.                                   }
VAR i, j, k, s, r, rowsout, colsout : integer;  min : real;
     rowin : boolcol;  colin : boolrow;
BEGIN rowsout:=0;  colsout:=0;    { Initialise local variables, }
   FOR i:=1 TO m DO rowin[i]:=true; { including row and column    }
   FOR j:=1 TO n DO colin[j]:=true; { elimination indicators.     }
   FOR k:=1 TO m+n-1 DO { Find basic variables }
   BEGIN min:=largevalue;
      FOR i:=1 TO m DO   { Search remaining rows in table }
        IF rowin[i] THEN
           FOR j:=1 TO n DO     { Search remaining columns }
             IF colin[j] THEN   { within remaining rows.   }
                IF (min>T[i,j].c) THEN BEGIN min:=T[i,j].c; s:=i; r:=j END;
      T[s,r].basic := true;  { New basic variable is at position (s,r) }
      basrowno[s] := basrowno[s] + 1;  bascolno[r] := bascolno[r] + 1;
      { Find value of basic variable and update row or column value }
      IF (a[s]>b[r]) OR ((a[s]=b[r]) AND (rowsout=m-1)) THEN
      BEGIN  T[s,r].x:=b[r];  a[s]:=a[s]-b[r];
             colin[r]:=false;  colsout:=colsout+1
      END
      ELSE
      BEGIN  T[s,r].x:=a[s];  b[r]:=b[r]-a[s];
             rowin[s]:=false;  rowsout:=rowsout+1
      END;
      totalcost := totalcost + T[s,r].x*T[s,r].c
   END
END; { firstbfs }


PROCEDURE newbfs;   { Find the next basis feasible solution }
VAR  i, j, newvarrow, newvarcol, oldvarrow, oldvarcol : integer;
     diff, cdashij, w : real;  u : col;  v : row;

   PROCEDURE findusandvs; { Find the row and column Simplex multipliers }
   VAR  i, maxbasno, sstart : integer;
        uset : boolcol;  vset : boolrow;
```

```
PROCEDURE initialise;
VAR  i, j : integer;
BEGIN  maxbasno:=0;
   FOR i:=1 TO m DO uset[i]:=false;   FOR j:=1 TO n DO vset[j]:=false
END; { initialise }

PROCEDURE colsearch (r:integer); FORWARD;

PROCEDURE rowsearch (s:integer);
VAR j : integer;
BEGIN
   IF basrowno[s]>0 THEN
      FOR j:=1 TO n DO
         IF T[s,j].basic AND NOT vset[j] THEN
         BEGIN  v[j] := T[s,j].c - u[s];
                vset[j]:=true;  colsearch(j)

         END
END; { rowsearch }               { rowsearch and colsearch are a pair of }
                                 { mutually recursive procedures which    }
PROCEDURE colsearch;             { search rows and columns of T for basic }
VAR  i : integer;                { variables in order compute Simplex     }
BEGIN                            { multipliers using equation (4.8).      }
   IF bascolno[r]>0 THEN
      FOR i:=1 TO m DO
         IF T[i,r].basic AND NOT uset[i] THEN
         BEGIN  u[i] := T[i,r].c - v[r];
                uset[i]:=true;  rowsearch(i)
         END
END; { colsearch }

BEGIN { findusandvs }
  initialise;
  FOR i:=1 TO m DO { Find the row containing most basic variables }
    IF basrowno[i] > maxbasno THEN
      BEGIN  maxbasno := basrowno[i];  sstart:=i  END;
  u[sstart]:=0;  uset[sstart]:=true;  rowsearch(sstart)
END; { findvsandus }

PROCEDURE modifytable;
{ Find the value, w, of the new basic variable and that basic   }
{ variable which should leave the basis. Modify any variable    }
{ values in order to preserve row and column totals. This is    }
{ done through procedures rowpath and colpath below.            }
VAR  inpath : ARRAY [mrange,nrange] OF boolean;
     removed, pathfound : boolean;

  PROCEDURE initialise;
  VAR  i, j : integer;
  BEGIN  removed:=false;  pathfound:=false;
     FOR i:=1 TO m DO FOR j:=1 TO n DO inpath[i,j]:=false
  END; { initialise }

  PROCEDURE colpath (s,r:integer; y:real); FORWARD;
```

```pascal
    PROCEDURE rowpath (s,r:integer; y:real);
    VAR  j :integer;   z : real;
    BEGIN  inpath[s,r]:=true;  j:=0;
      WHILE (j<n) AND NOT pathfound DO
      BEGIN  j:=j+1;
        IF T[s,j].basic THEN
          IF NOT inpath[s,j] THEN
            IF j<>newvarcol THEN
            BEGIN z:=T[s,j].x; IF y<z THEN z:=y;
              IF bascolno[j]>1 THEN colpath(s,j,z)
            END
            ELSE
            BEGIN  pathfound:=true;
              IF T[s,j].x<y THEN w:=T[s,j].x ELSE w:=y;
              IF printon THEN BEGIN printval(w); writeln END
            END
      END;
      IF NOT pathfound THEN inpath[s,r]:=false  { Row is unproductive }
      ELSE
      BEGIN  T[s,j].x := T[s,j].x - w;  { Update basic variable value }
        IF (T[s,j].x < smallvalue) AND NOT removed THEN
        BEGIN IF printon THEN writeln('     X[', s:fwi, ',', j:fwi, ']',
                             '  LEAVES THE BASIS'); removed:=true;
          T[s,j].basic:=false; oldvarrow:=s; oldvarcol:=j;
        END
      END
    END; { rowpath }

    PROCEDURE colpath;
    VAR  i : integer;   z : real;
    BEGIN  inpath[s,r]:=true;  i:=0;
      WHILE (i<m) AND NOT pathfound DO
      BEGIN  i:=i+1;
        IF T[i,r].basic THEN
          IF NOT inpath[i,r] THEN
            IF i<>newvarrow THEN
            BEGIN IF a[i]<b[r] THEN z:=a[i] ELSE z:=b[r];
              z:=z-T[i,r].x; IF y<z THEN z:=y;
              IF basrowno[i]>1 THEN rowpath(i,r,z)
            END
      END;
      IF NOT pathfound THEN inpath[s,r]:=false { Column is unproductive }
      ELSE
      BEGIN T[i,r].x := T[i,r].x + w END  { Update basic variable value }
    END; { colpath }
```

{ rowpath and colpath are a pair }
{ of mutually recursive }
{ procedures which search rows }
{ and columns of T in order to }
{ find the value, w, of the new }
{ basic variable. They also }
{ alter values of other basic }
{ variables so that row and }
{ column totals are preserved. }

```pascal
BEGIN { modifytable }
  initialise;
  IF a[newvarrow] < b[newvarcol] THEN w:=a[newvarrow]
  ELSE w:=b[newvarcol];
  rowpath(newvarrow,newvarcol,w); T[newvarrow,newvarcol].x:=w
END; { modifytable }
```

```
BEGIN { newbfs }
  findusandvs;
  cdashij:=largevalue;
  FOR i:=1 TO m DO     { Find smallest c´ij }
    FOR j:=1 TO n DO
      IF NOT T[i,j].basic THEN
      BEGIN  diff := T[i,j].c - u[i] - v[j];
        IF diff < cdashij THEN
        BEGIN cdashij:=diff; newvarrow:=i;  newvarcol:=j  END
      END;
    solution := cdashij > -smallvalue;  { Solution found if all }
    IF NOT solution THEN               { c´ij are positive.    }
    BEGIN
      IF printon THEN
      BEGIN  writeln(´    SHADOW COSTS´);  write(´    U[I]   ´);
        FOR i:=1 TO m DO printval(u[i]);  writeln;  write(´    V[J]   ´);
        FOR j:=1 TO n DO printval(v[j]);  writeln;
        write(´    X[´, newvarrow:fwi, ´,´, newvarcol:fwi, ´] ENTERS THE´,
              ´ BASIS WITH C´´[´, newvarrow:fwi, ´,´, newvarcol:fwi,
              ´] =´); printval(cdashij); write(´  AND W =´)
      END;
      T[newvarrow,newvarcol].basic := true;
      basrowno[newvarrow] := basrowno[newvarrow] + 1;
      bascolno[newvarcol] := bascolno[newvarcol] + 1;
      modifytable;
      totalcost := totalcost + cdashij*T[newvarrow,newvarcol].x;
      basrowno[oldvarrow] := basrowno[oldvarrow] - 1;
      bascolno[oldvarcol] := bascolno[oldvarcol] - 1;
    END
END; { newbfs }

BEGIN  { Main Program }
  writeln; writeln(´´´STEPPING STONES´´ ALGORITHM´); writeln;
  inputdata;  initialise;  outputtable;  firstbfs(a,b);
  REPEAT
    IF printon THEN outputbfs;  it:=it+1;
    newbfs
  UNTIL solution;  IF NOT printon THEN outputbfs;  writeln;
  writeln(´    MINIMISATION ACHIEVED - FINAL SOLUTION IS GIVEN ABOVE´)
END.
```

´STEPPING STONES´ ALGORITHM

    INITIAL TABLE

J	1	2	3	4	5	A[I]
I						
1	1	0	3	4	2	15
2	5	1	2	3	3	25
3	4	8	1	4	3	20
B[J]	20	12	5	8	15	

```
BASIC FEASIBLE SOLUTION NUMBER 0
BASIC VARIABLES  I  J      X[I,J]        C[I,J]           COST
                 1  1         3             1              3
                 1  2        12             0              0
                 2  1         2             5             10
                 2  4         8             3             24
                 2  5        15             3             45
                 3  1        15             4             60
                 3  3         5             1              5
                                          TOTAL COST =    147
SHADOW COSTS
U[I]      -4     0    -1
V[J]       5     4     2     3     3
X[2,2] ENTERS THE BASIS WITH C´[2,2] =    -3  AND W =      2
X[2,1]  LEAVES THE BASIS

BASIC FEASIBLE SOLUTION NUMBER 1
BASIC VARIABLES  I  J      X[I,J]        C[I,J]           COST
                 1  1         5             1              5
                 1  2        10             0              0
                 2  2         2             1              2
                 2  4         8             3             24
                 2  5        15             3             45
                 3  1        15             4             60
                 3  3         5             1              5
                                          TOTAL COST =    141
SHADOW COSTS
U[I]      -1     0     2
V[J]       2     1    -1     3     3
X[3,5] ENTERS THE BASIS WITH C´[3,5] =    -2  AND W =     10
X[1,2]  LEAVES THE BASIS

BASIC FEASIBLE SOLUTION NUMBER 2
BASIC VARIABLES  I  J      X[I,J]        C[I,J]           COST
                 1  1        15             1             15
                 2  2        12             1             12
                 2  4         8             3             24
                 2  5         5             3             15
                 3  1         5             4             20
                 3  3         5             1              5
                 3  5        10             3             30
                                          TOTAL COST =    121

MINIMISATION ACHIEVED - FINAL SOLUTION IS GIVEN ABOVE
```

Exercises 4

1 Solve the following transportation problem

4	3	3	1	8
3	2	4	8	11
5	4	6	3	16
4	9	9	13	

2 Solve the following transportation problem

2	9	3	10	8	7	9
3	2	6	3	5	2	14
1	8	2	4	1	5	16
4	8	7	6	6	6	11
6	4	10	13	7	10	

3 A steel company has three mills M_1, M_2 and M_3 capable of producing 50, 30 and 20 thousand tonnes of steel respectively over a certain period. The company supplies four customers C_1, C_2, C_3 and C_4 whose respective requirements are for 12, 15, 25 and 36 thousand tonnes of steel. The cost of producing and transporting one thousand tonnes of steel from the various mills to the various customers are shown below. Determine the production at each mill and the transportation pattern that will minimise the total cost.

	Mills		
	M_1	M_2	M_3
C_1	15	19	14
C_2	19	18	16
C_3	19	18	20
C_4	15	19	18

Customers

4   A company is planning to move many of its executive staff to different jobs following an establishment review in which several new posts were created and some old ones discontinued. The staff affected may be divided, according to their qualifications and experience, into five groups S_1, S_2, S_3, S_4 and S_5 containing 2, 5, 4, 8 and 6 staff respectively. The posts may be similarly split into four groups P_1, P_2, P_3 and P_4 of 8, 3, 9 and 5 jobs respectively. The table below shows by an asterisk which groups of staff possess the right backgrounds for the various types of post.

		Staff				
		S_1	S_2	S_3	S_4	S_5
	P_1		*		*	
Post	P_2				*	*
	P_3	*			*	
	P_4	*	*	*		*

Use the transportation method to show that no entirely satisfactory deployment of staff is possible and determine the maximum number of staff that can be allocated suitable jobs. Would it be possible to restrict all the unsatisfactory allocations to the staff in group S_5?

5   A company controls three factories F_1, F_2 and F_3 capable of producing 50, 25 and 25 thousand components respectively per week. The company has contracted to supply four customers C_1, C_2, C_3 and C_4 who require 15, 20, 20 and 30 thousand components respectively per week. The costs of producing and transporting one thousand components from the factories to the customers are shown below. Determine the production at each factory and the distribution pattern which will minimise the total cost.

		Customer			
		C_1	C_2	C_3	C_4
	F_1	13	17	17	14
Factory	F_2	18	16	16	18
	F_3	12	14	19	17

6   A company owns two factories F_1 and F_2 which manufacture electronic components. The production capacities of the two factories F_1 and F_2 over a certain period are 16 000 and 12 000 components respectively during normal time but if overtime is worked this can be raised to 20 000 and 14 000 components respectively. The additional cost of producing 1000 components in overtime working at both F_1 and F_2 is 8 units. The company supplies three customers C_1, C_2 and C_3 whose requirements over the same period are for 10 000, 13 000 and 7000 components respectively. The costs of transporting 1000 components from factories to customers are given on the next page.

	To		
	C_1	C_2	C_3
From F_1	5	4	6
F_2	6	3	2

Formulate the problem of finding the optimum schedule and the pattern of distribution from the factories to the customers as a transportation problem and find a solution.

7   A large firm has invited three airlines to send in their tenders for flying a team of the company's technicians to various parts of the world as required. The tenders in £100's are shown below.

		Sydney	Calcutta	Beirut	Dallas	Sao Paulo
	I	24	16	8	10	14
Airline	II	21	15	7	12	16
	III	23	14	7	14	12

The directors have decided that the individual flight contracts are to be allocated in the ratio $2:3:2$ between airlines I, II and III respectively and inform the transport manager of this. He also knows that of the 70 flights next year 10 will be to Sydney, 15 to Calcutta, 20 to Beirut, 10 to Dallas and 15 to Sao Paulo.

How must he allocate the individual flight contracts in order to minimise the total cost subject to the directors' orders? What is the minimum cost of providing the required transport?

8   Four steel works I, II, III, IV produce respectively 950, 300, 1350 and 450 tonnes of a particular grade of steel per week. The ingots have to be conveyed to consumers A, B, C, D, E whose weekly requirements are for 250, 1000, 700, 650 and 450 tonnes of steel.

The cost of transportation per tonne, in appropriate units, from the works to the consumers is given below.

		Consumer				
		A	B	C	D	E
	I	12	16	21	19	32
Works	II	4	4	9	5	24
	III	3	8	14	10	26
	IV	24	33	36	34	49

Find the allocation pattern which minimises the costs of distribution.

9   One product is made at two different plants and shipped to two consumers. Their requirements for the next two months are for the following quantities.

	Required for	
	August	September
Consumer I	420	550
Consumer II	350	480

Unit shipping costs from plants to consumers are given in the table below.

From/To	Consumer I	Consumer II
Plant I	10	13
Plant II	12	6

Manufacturing costs per unit, together with the production capacities of the plants, for the next two months are

	Production costs		Capacity	
	August	September	August	September
Plant I	3.0	3.6	500	600
Plant II	3.2	2.9	300	500

It is possible to manufacture units during one month, store them for a month and then ship them to the consumers. Storage costs per unit per month are 0.5 at Plant I and 0.6 at Plant II.

   The optimum production schedule and distribution pattern is required. Formulate this problem as a transportation problem and hence find an optimal solution.

10   A company operates three plants A, B and C. Their respective production costs are 26c, 23c and 22c per unit and their capacities are 6000, 3000 and 3000 units. The company has contracted to supply 1500, 2500, 2700 and 3300 units respectively at cities W, X, Y, Z. Given the transportation costs below draw up the optimum production and distribution schedule.

		Transportation cost (cents per unit) From		
		A	B	C
To	W	1	9	6
	X	4	2	1
	Y	1	2	7
	Z	9	8	3

11 Solve the problems of this chapter and these exercises using the program for the Simplex Method. Does it appear to be inefficient as compared with the transportation program?

12 A first basic feasible solution in the transportation problem can be found using the 'North-West corner' rule. Here the cell to which the next basic variable is assigned is the one in the north-west corner of the remaining array, irrespective of the cost in this cell. Thus the rows are systematically deleted from the top and the columns from the left.

Modify the program to choose the first basic feasible solution in this way. Solve the problems of this chapter and these exercises with your program. Do you notice any gain or loss of efficiency?

5
The Assignment Problem

5.1 Introduction

Example 1

Five men $M_1, M_2, \ldots, M_5$ are capable of doing 5 tasks $T_1, T_2, \ldots, T_5$. Because of their differing aptitudes the men will take different times to complete the different tasks. These times are given in hours in the table below. How should the men be assigned to the tasks in order to minimise the total man hours to complete the tasks?

	T_1	T_2	T_3	T_4	T_5
M_1	10	5	9	18	11
M_2	13	19	6	12	14
M_3	3	2	4	4	5
M_4	18	9	12	17	15
M_5	11	6	14	19	10

Let x_{ij} be the proportion of the ith man assigned to the jth job. The x_{ij} are non-negative and since each man is to be fully utilised and each job is to be fully manned, the x_{ij} must satisfy the constraints:

$$x_{11} + x_{12} + \cdots \qquad + x_{15} = 1$$
$$\cdots \cdots \cdots \cdots \cdots \cdots \cdots \cdots \cdots \cdots$$
$$x_{51} + x_{52} + \cdots \qquad + x_{55} = 1$$
$$x_{11} + x_{21} + \cdots \qquad + x_{51} = 1$$
$$\cdots \cdots \cdots \cdots \cdots \cdots \cdots \cdots \cdots \cdots$$
$$x_{15} + x_{25} + \cdots \qquad + x_{55} = 1.$$

Subject to these constraints we seek to minimise the total time:

$$T = 10x_{11} + 5x_{12} + \cdots \qquad + 19x_{54} + 10x_{55}.$$

Thus the problem is a linear programming problem of the transportation type. Since all row and column totals are 1, it will be highly degenerate, so that in general the transportation algorithm will not be an efficient solution procedure, although it would work. Since it is a transportation problem, in the optimal solution, which will be integral, 5 of the x_{ij}'s will be 1, the rest 0. Thus another way of looking at the problem is that from the 5×5 matrix of times we have to choose 5 elements, one in each row and one in each column such that the sum of the elements chosen is a minimum. This 0, 1 requirement on the x_{ij} is necessary for our formulation to make sense in some contexts (the first sentence used in the formulation of this problem for example) but it is assured.

The problem can be generalised to an $n \times n$ matrix. From such a matrix the problem is to choose n elements, one in each row and one in each column so that the sum of these elements is minimised. The chosen elements will be $x_{1i}, x_{2j}, ..., x_{nt}$ where $i, j, ..., t$ is some permutation of $1, 2, ..., n$. There are thus $n!$ ways of choosing the elements so that for even moderate values of n a solution by complete enumeration becomes prohibitively long.

5.2   Mack's Method of Solution

To date two methods for solving the assignment problem have been proposed. Both are based on the fact that the *positions* of the optimal assignment are unaltered if to each element in a row or in a column, we add or subtract the same quantity.

The Hungarian method is based on some rather difficult and obscure combinatorial properties of matrices. It is not easy to program and we shall content ourselves with merely stating that a description of the procedure is to be found in many books on Operational Research and mathematical programming.

Mack's Bradford method has the advantage that it has a simpler intuitive basis. It is a logical step by step process. We shall describe the steps in this iterative process and transform them into a computer program in Section 5.3. It is based on the idea of choosing the element in each row to be the minimum element in the row. In general the row minima are not spread over the n columns of the matrix. We use the idea of adding or subtracting the same quantity to each element in a row or a column to spread the row minima over the columns, one to a column. They will then constitute a minimum assignment.

Mack's Bradford Method for the Assignment Problem

The originator of the method, C. Mack, gives a very good account in the *New Journal of Statistics and Operational Research,* Vol. 1, p. 17.

We consider an $n \times n$ assignment problem.

Choose the minimum element of each row. Underline each of these minimum elements. If we have one underlined element in each column the underlined elements or **bases** give the optimal allocation.

START

Divide the columns into two sets A and A', A the selected set and A' the unselected set. Initially, and at subsequent returns to START, we have not selected any column so A is empty and A' contains all the columns.

Step 1 Select from A' a column containing more than one base. Transfer this column from A' to A.

Step 2 Let any base of A which is in row i have value b_i, and the minimum element of A' in row i have value a_i'.

Let
$$\min_i (a_i' - b_i) = a_r' - b_r.$$

Step 3 Increase every element of A by $a_r' - b_r$.
Step 4 Underline a_r' with dots. a_r' is now a 'dotted element'.
Step 5 Let the column containing a_r' be C. If C already contains one or more bases remove C from A' to A and go to step 2. Otherwise go to the next step.
  We are now ready to spread the bases over one more column.
Step 6 Underline fully the last element a_r' selected. This is now a new base.
Step 7 Find the orginal base in the same row as a_r'. Remove the underlining. Call the column of this base column D.
Step 8 If D does not contain a further base it must contain a dotted element. Call this dotted element a_r' and go back to step 6.
  If D does contain a further base the elements now fully underlined form the new set of bases.
  If we still have any column without a base go back to START.
  If every column contains a base we have finished. The positions of the optimal bases can now be plotted on the original allocation and the optimum cost or return calculated.
  The computations may be shortened slightly if at step 3 we increase only the bases of A by $a_r - b_r$, waiting until step 8 before increasing the values of the other elements of A. In this case all remaining elements of a column are raised by the same amount as any base in the column has been raised.
  A full description of the method as applied to Example 1 of Section 5.1 follows. The reader should follow through the working carefully.

START
Step 1 Col. 2 goes to A   10   5   9   18   11
Step 2 Row 1, base 5, min. in A' is   9, difference = 4   13   19   6   12   14
       Row 3, base 2, min. in A' is   3, difference = 1   3   2   4   4   5
       Row 4, base 9, min. in A' is 12, difference = 3   18   9   12   17   15
       Row 5, base 6, min. in A' is 10, difference = 4   11   6   14   19   10
Minimum of $(a_i' - b_i) = 1$ in row 3.
Step 3 Increase every element in column 2 by 1   10   6   9   18   11
Step 4 Dot underline 3 in row 3 col. 1   13   20   6   12   14
Step 5 C (col. 1) contains no other bases so go to   3   3   4   4   5
Step 6 Fully underline 3 in row 3 col. 1   18   10   12   17   15
Step 7 Remove underlining from 3 in row 3 col. 2   11   7   14   19   10
Step 8 D (col. 2) contains other bases so return to
START since we have columns without bases.

START
Step 1   Col. 2 goes to A
Step 2   Min $(a_i^j - b_i)$ is $12 - 10 = 2$ in row 4
Step 3   Increase every element in col. 2 by 2
Step 4   Dot underline 12 in row 4 col. 3
Step 5   C (col. 3) contains another base so transfer col. 3 to A and go to step 2
Step 2   A is now col. 2 and col. 3. Minimum of $a_i^j - b_i$ is $10 - 9 = 1$ in row of 5

```
10   8    9   18   11
13  22    6   12   14
 3   5    4    4    5
18  12   12   17   15
11   9   14   19   10
```

Step 3   Increase every element in col. 2 and col. 3 by 1
Step 4   Dot underline 10 in row 5 col. 5
Step 5   C (col. 5) contains no other bases so go to
Step 6   Fully underline 10 in row 5 col. 5
Step 7   Remove underlining from 10 in row 5 col. 2
Step 8   D (col. 2) contains other bases so go to
START since there is no base in col. 4.

```
10   9   10   18   11
13  23    7   12   14
 3   6    5    4    5
18  13   13   17   15
11  10   15   19   10
```

START
Step 1   Col. 2 is A
Step 2   Minimum $(a_i^j - b_i) = 13 - 13 = 0$
Step 3   Changes nothing
Step 4   Dot underline 13 in row 4 col. 3
Step 5   C is col. 3 and contains another base. Go to
Step 2   Col. 2 and col. 3 are in A.
   Min $(a_i^j - b_i) = 10 - 9 = 1$

```
10   9   10   18   11
13  23    7   12   14
 3   6    5    4    5
18  13   13   17   15
11  10   15   19   10
```

Step 3   Increase cols. 2 and 3 by 1
Step 4   Dot underline 10 in row 1 col. 1
Step 5   C (col. 1) contains another base so go to
Step 2   A is now col. 1, col. 2 and col. 3
   Min $(a_i^j - b_i) = 11 - 10 = 4 - 3 = 15 - 14 = 1$

```
10  10   11   18   11
13  24    8   12   14
 3   7    6    4    5
18  14   14   17   15
11  11   16   19   10
```

Step 3   Increase cols. 1, 2 and 3 by 1
Step 4   Dot underline 4 in row 3 col. 4
Step 5   C (col. 4) contains no other base so go to
Step 6   Fully underline 4 in row 3 col. 4
Step 7   Remove underlining from 4 in row 3 col. 1
Step 8   D (col. 1) contains a dotted element in row 1.
   So go to
Step 6   Fully underline 11 in row 1 col. 1
Step 7   Remove underlining from 11 in row 1 col. 2
Step 8   D (col. 2) contains another base and the new set of bases is as shown.

```
11  11   12   18   11
14  25    9   12   14
 4   8    7    4    5
19  15   15   17   15
12  12   17   19   10
```

We have a base in every column so this gives the final solution
Man 1 does job 1, man 2 does job 3, man 3 does job 4, man 4 does job 2, man 5 does job 5.

```
11  11   12   18   11
14  25    9   12   14
 4   8    7    4    5
19  15   15   17   15
12  12   17   19   10
```

Thus in terms of the original array, allocation and times are:

$$\begin{array}{ccccc}
\underline{10} & 5 & 9 & 18 & 11 \\
13 & 19 & \underline{6} & 12 & 14 \\
3 & 2 & 4 & \underline{4} & 5 \\
18 & \underline{9} & 12 & 17 & 15 \\
11 & 6 & 14 & 19 & \underline{10}
\end{array}$$

Minimum time taken is 39 man hours.

Example 2

Each day an airline operates the following flights between two cities X and Y.

Flight No.	Dep. X	Arr. Y	Flight No.	Dep. Y	Arr. X
1	9.00	11.00	11	8.00	10.00
2	10.00	12.00	12	9.00	11.00
3	15.00	17.00	13	14.00	16.00
4	19.00	21.00	14	20.00	22.00
5	20.00	22.00	15	21.00	23.00

The company wishes to pair the arriving flights with the departing flights at both X and Y so as to minimise time on the ground (subject to this being at least one hour to allow for refuelling). Use the assignment technique to carry out this pairing.

Hence write down the sequence of flights undertaken by any of the aeroplanes. Represent the solution on a diagram. How many aeroplanes are needed to operate the schedule?

The arrays below show the time on the ground for each possible pairing.

At X In/Out	1	2	3	4	5	At Y In/Out	11	12	13	14	15
11	23	24	5	9	10^*	1	21	22	3	9	10^*
12	22	23	4^*	8	9	2	20	21	2^*	8	9
13	17	18	23	3^*	4	3	15	16	21	3^*	4
14	11^*	12	17	21	22	4	11^*	12	17	23	24
15	10	11^*	16	20	21	5	10	11^*	16	22	23

Thus we have two assignment problems to solve. Mack's method easily shows that the asterisked elements consitute a solution (although there is another). The reader should verify the solution.

Thus at X, pairings are 11—5, 12—3, 13—4, 14—1, 15—2;
and at Y, pairings are 1—15, 2—13, 3—14, 4—11, 5—12;
so sequence is

$$1—15—2—13—4—11—5—12—3—14—1 \text{ etc.}$$

Thus the aeroplane that leaves X at 9.00 on Monday completes all flights in the

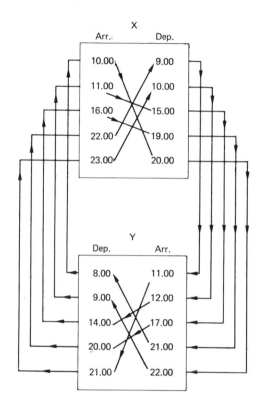

schedule by 22.00 on Thursday and is ready to start the sequence again at 9.00 on Friday. Thus 4 aeroplanes are needed to operate the schedule.

5.3   A Computer Program for Mack's Method

The computer program given in this section follows the iterative process just described. It is not trivial and requires careful study. It is written to maximise or minimise as required, this being determined by the first value input. Comments in the program indicate where the various parts of the algorithm are being implemented. The original $n \times n$ matrix of values, R, is preserved and the computation carried out on a copy, M. Linked lists of integers are used to hold some of the required information, namely the columns of M that are in A or A' and the rows of the bases in each column. This device makes for easy access to the current bases and easy transfer of bases from one column to another. Movement of columns from A to A' and of bases between columns is effected by the PROCEDURE *move*.

The Main Program part begins with data input and output and with required initialisations. The initial set of bases is then found before entering the algorithm proper. This is contained within a WHILE loop, checking each time to see if a solution (one base in each row and column) has been found, and using a FUNCTION *finished* for this.

The various steps are clearly marked in the program. Steps 2 to 5 are contained in a REPEAT loop, the majority of which is concerned with Step 2. In this step we are seeking the row (stored in *rowmin*) which leads to the smallest increase in array elements should there be a transfer of bases within columns of this row. This search is for old bases through the columns of *A* (using the pointer *p* to identify columns and *q* to find the row of each base). Potential new bases are found by identifying each column in *A'* using pointer *pdash* and testing for the minimum difference between old base and new base (this latter being stored in *colmin*).

The modified Step 3 is then used to store the new 'column raised' amount and Step 4 makes the potential new base a 'dotted element'. If this column (now called *C* to conform with the notation of the algorithm) has no bases we can proceed to distribute a base into this column (Step 6), but not if we need to continue our search at Step 2. Step 5 and the UNTIL part of the REPEAT loop decide which course of action to take.

Eventually Step 6 will be reached, placing the new base in column *C* and then Step 7 removes the old base at column *D* (remembered from Step 4). The *noofbase* counters are altered appropriately. We then reach Step 8 and if column *D* has a base left in we can return to the head of the main WHILE loop to check for a solution. Otherwise the dotted element in column *D* will be made a new base and an old one removed by repeating Steps 6 and 7. The solution and the minimum are output.

Example 1

For the matrix

$$
\begin{array}{cccccccc}
93 & 93 & 91 & 94 & 99 & 99 & 90 & 92 \\
96 & 93 & 90 & 94 & 98 & 96 & 97 & 91 \\
96 & 90 & 91 & 90 & 92 & 90 & 93 & 96 \\
93 & 94 & 95 & 96 & 97 & 10 & 92 & 93 \\
94 & 93 & 95 & 91 & 90 & 97 & 96 & 92 \\
94 & 93 & 96 & 90 & 93 & 89 & 88 & 91 \\
94 & 96 & 91 & 90 & 95 & 93 & 92 & 94 \\
93 & 94 & 6 & 95 & 91 & 99 & 91 & 96
\end{array}
$$

find (i) a minimal, (ii) a maximal assignment. The data above, preceded by -1 for case (i) and $+1$ for case (ii) led to the output as shown following the program listing.

```
PROGRAM Mack (input,output);
CONST
  n=8;               { Matrix size }                                    {**}
  fwx=7; dpx=1;  { Output format constants for matrix values }          {**}
  largevalue=1.0E20;                                                     {**}

TYPE
  nrange = 1..n;
  matrix = ARRAY [nrange,nrange] OF real;
  numlist = ^numnode;         { Structure for linked lists of integers. }
  numnode = RECORD            { Will be used to store indices of matrix }
              num : integer; { columns of A and A´ as well as row       }
              next : numlist { indices of bases in each matrix column.   }
            END;
```

```
        colinfo = RECORD  { Stores information for each matrix column }
                   raise : real;  { Amount to which column is to be raised }
                   dotted : integer;  { Index of dotted element, if any }
                   noofbases : integer;  { No. of bases in column }
                   baselist : numlist  { List of row indices of bases }
                END;

    colrec = ARRAY [nrange] OF colinfo;
    baserec = ARRAY [nrange] OF integer;


VAR
   R, M : matrix;        { Original and working matrices }
   column : colrec;      { Information for all columns }
   oldbase : baserec;    { Previous base in given row }
   A, Adash : numlist;   { Index lists for A and A' }
   rows : numlist;  maximise : boolean;
   i, j, k, col, baserow : integer;
   min, bi, aidash, sum : real;
   colmin, rowmin,
   oldbasecol, C, D : integer;
   p, q, pdash : numlist;


PROCEDURE inputdata;
VAR  i, j, k : integer;
BEGIN  read(k);  maximise := k>0;
  FOR i:=1 TO n DO FOR j:=1 TO n DO read(R[i,j])
END; { inputdata }


PROCEDURE outputdata;
VAR  i, j : integer;
BEGIN writeln('    ', n:3, ' BY', n:3, ' MATRIX OF VALUES'); writeln;
  FOR i:=1 TO n DO
  BEGIN FOR j:=1 TO n DO write(R[i,j]:fwx:dpx); writeln END
END; { outputdata }


PROCEDURE initialise;
VAR  i, j : integer;
  FUNCTION fulllist: numlist;
  { Produces a linked list containing integers 1 through n }
  VAR  x, y : numlist;  i : integer;
  BEGIN  new(x);  x^.num:=n;  x^.next:=nil;
    FOR i:=n-1 DOWNTO 1 DO { insert new node on front of list }
    BEGIN  new(y);  y^.num:=i;  y^.next:=x;  x:=y  END;
    fulllist := x
  END; { fulllist }
BEGIN { initialise }
  A:=nil;  Adash:=fulllist;  rows:=fulllist;
  FOR i:=1 TO n DO WITH column[i] DO
  BEGIN  dotted:=0;  noofbases:=0;  baselist:=nil  END;
  M:=R;
  IF maximise THEN FOR i:=1 TO n DO FOR j:=1 TO n DO M[i,j]:=-M[i,j]
END; { initialise }
```

```
PROCEDURE move (k:integer; VAR lista, listb : numlist);
{ Move the integer k from lista to listb }
VAR   p, q : numlist;
BEGIN   p:=lista;
  IF p^.num = k THEN lista := lista^.next
  ELSE
  BEGIN REPEAT q:=p; p:=p^.next UNTIL p^.num=k; q^.next:=p^.next END;
  p^.next := listb;  listb:=p
END; { move }




FUNCTION finished: boolean;
{ Checks whether there is a single base in each column of the matrix }
VAR  b : boolean; i : integer;
BEGIN  b:=true;  i:=1;
  WHILE b AND (i<=n) DO
  BEGIN  b := column[i].noofbases=1;  i := i+1   END;
  finished := b
END; { finished }




BEGIN   { Main Program }
  writeln;
  writeln('     ASSIGNMENT PROBLEM - MACK''S METHOD OF SOLUTION');
  writeln;  inputdata;  outputdata;   initialise;
  { Compute initial bases and column information }
  FOR i:=1 TO n DO { Find base in each row of matrix }
  BEGIN   min:=M[i,1];  col:=1;
    FOR j:=2 TO n DO
      IF M[i,j]<min THEN BEGIN min:=M[i,j]; col:=j END;
    move(i, rows, column[col].baselist);
    column[col].noofbases := column[col].noofbases + 1
  END;




  WHILE   NOT finished  DO    { Check for solution }
  BEGIN
    { START }
    WHILE A<>nil DO move(A^.num, A, Adash); { Empty A and fill A' }
    FOR i:=1 TO n DO column[i].raise:=0.0;
    { Step 1 }  k:=1;  WHILE column[k].noofbases<2 DO k:=k+1;
    move(k, Adash, A); { Column k is first to enter A from A' }
    REPEAT   { Until a column without a base is encountered }
      { Step 2 }  min:=largevalue;  p:=A;
      WHILE p<>nil DO { Search all columns of A for bases }
      BEGIN   col:=p^.num;  q:=column[col].baselist;
        WHILE q<>nil DO { Search current column for bases }
        BEGIN   baserow:=q^.num;
          bi := M[baserow,col] + column[col].raise;  { True base value }
          pdash:=Adash;  aidash:=largevalue;
          WHILE pdash<>nil DO { Search row of A' for minimum }
```

```
            BEGIN   j:=pdash^.num;
              IF M[baserow,j]<aidash THEN
                BEGIN aidash:=M[baserow,j]; k:=j END;
              pdash:=pdash^.next
            END;
            IF (aidash-bi) < min THEN
            BEGIN min:=aidash-bi; colmin:=k; rowmin:=q^.num;
                  oldbasecol:=col
            END;
            q:=q^.next
          END;
        p:=p^.next
      END;

      { Modified Step 3 }   p:=A;
      WHILE p<>nil DO { Amend column raise values in A }
      BEGIN   col:=p^.num;   column[col].raise := column[col].raise + min;
              p:=p^.next
      END;
      { Step 4 }   column[colmin].dotted:=rowmin;
                   oldbase[rowmin]:=oldbasecol;
      { Step 5 }   C:=colmin;
                   IF column[C].noofbases > 0 THEN move(C, Adash, A)
    UNTIL column[C].noofbases = 0;



    REPEAT   { Until a column containing at least one base is found }
      { Steps 6 & 7 }   D:=oldbase[rowmin];
      { Dotted element in M[rowmin,C] is fully underlined and old }
      { underlined element in M[rowmin,D] has underline removed   }
      move(rowmin, column[D].baselist, column[C].baselist);
      column[D].noofbases := column[D].noofbases - 1;
      column[C].noofbases := column[C].noofbases + 1;
      column[C].dotted:=0;
      { Step 8 }
      IF column[D].noofbases = 0 THEN
      BEGIN   rowmin := column[D].dotted;   C:=D   END
    UNTIL column[D].noofbases > 0
  END;



  { Output solution }   writeln;
  writeln('   SOLUTION - GIVEN IN ASCENDING COLUMN ORDER');   writeln;
  writeln('     I  J', '  ':fwx-6, 'R[I,J]');   writeln;   sum:=0;
  FOR j:=1 TO n DO
  BEGIN   i := column[j].baselist^.num;   sum := sum + R[i,j];
          writeln('    ', i:3, j:3, R[i,j]:fwx:dpx)
  END;   writeln;
  IF maximise THEN write('   MAXIMUM') ELSE write('   MINIMUM');
  writeln(' IS', sum:fwx:dpx)
END. { Mack }
```

ASSIGNMENT PROBLEM – MACK´S METHOD OF SOLUTION

8 BY   8 MATRIX OF VALUES

93.0	93.0	91.0	94.0	99.0	99.0	90.0	92.0
96.0	93.0	90.0	94.0	98.0	96.0	97.0	91.0
96.0	90.0	91.0	90.0	92.0	90.0	93.0	96.0
93.0	94.0	95.0	96.0	97.0	10.0	92.0	93.0
94.0	93.0	95.0	91.0	90.0	97.0	96.0	92.0
94.0	93.0	96.0	90.0	93.0	89.0	88.0	91.0
94.0	96.0	91.0	90.0	95.0	93.0	92.0	94.0
93.0	94.0	6.0	95.0	91.0	99.0	91.0	96.0

SOLUTION – GIVEN IN ASCENDING COLUMN ORDER

I	J	R[I,J]
1	1	93.0
3	2	90.0
8	3	6.0
7	4	90.0
5	5	90.0
4	6	10.0
6	7	88.0
2	8	91.0

MINIMUM IS   558.0

ASSIGNMENT PROBLEM – MACK´S METHOD OF SOLUTION

8 BY   8 MATRIX OF VALUES

93.0	93.0	91.0	94.0	99.0	99.0	90.0	92.0
96.0	93.0	90.0	94.0	98.0	96.0	97.0	91.0
96.0	90.0	91.0	90.0	92.0	90.0	93.0	96.0
93.0	94.0	95.0	96.0	97.0	10.0	92.0	93.0
94.0	93.0	95.0	91.0	90.0	97.0	96.0	92.0
94.0	93.0	96.0	90.0	93.0	89.0	88.0	91.0
94.0	96.0	91.0	90.0	95.0	93.0	92.0	94.0
93.0	94.0	6.0	95.0	91.0	99.0	91.0	96.0

SOLUTION – GIVEN IN ASCENDING COLUMN ORDER

I	J	R[I,J]
2	1	96.0
7	2	96.0
6	3	96.0
4	4	96.0
1	5	99.0
8	6	99.0
5	7	96.0
3	8	96.0

MAXIMUM IS   774.0

Exercises 5

1 Solve the minimum assignment problems:

(a)			(b)			
3	7	6	4	6	9	7
7	4	3	13	10	14	14
4	3	3	9	9	16	13
			12	10	12	10

2 Solve the minimum assignment problems:

(a)					(b)					
9	20	60	15	21	44	74	35	49	30	45
38	71	69	49	60	22	28	42	59	83	41
28	13	80	28	34	28	39	54	47	35	24
58	34	13	37	25	49	53	45	50	43	38
30	3	53	20	21	27	37	30	18	30	22
					70	27	21	32	31	9

3 On a particular day a haulage company has to pick up 5 loads at the points A, B, C, D, E and deliver them to the points a, b, c, d, e. The distances between the pick-up points and final destinations of the loads are shown in Table 1 below.

Table 1 Distance in miles

A—a	B—b	C—c	D—d	E—e
60	30	100	50	40

The firm has available 5 lorries of two types X and Y at the points S, T, U, V, W, the type of lorry being—type X at S, type Y at T, type X at U, type X at V and type Y at W. The type X lorries are newer and more versatile than type Y and have lower costs associated with them. The running costs per mile of the two types of lorry (fuel, insurance, maintenance, etc.) are shown in Table 2.

Table 2 Cost/mile (c)

Type	Empty	Loaded
X	20	40
Y	30	60

The distances of the initial positions of the lorries from the loading points are shown in Table 3.

Table 3   Distance in miles

	A	B	C	D	E
S	30	20	40	10	20
T	30	10	30	20	30
U	40	10	10	40	10
V	20	20	40	20	30
W	30	20	10	30	40

Determine the allocation of the lorries to the loads which minimises the costs. It should be assumed that all the loads are of about the same size and will require the same amount of packing, handling, etc.

4   In a radar system designed to track aircraft automatically, calculations are performed to determine the relative likelihoods of each plot arising from each aircraft being tracked, giving data such as that in the table below.

		Aircraft			
		1	2	3	4
Plot	1	0·79	0·20	0·50	0·315
	2	0·63	0·40	0·20	0·50
	3	0·40	0·20	0·16	0·50
	4	0·50	0·20	0·125	0·25

Explain how methods of solving the assignment problem could be used in such a system to determine how plots should be assigned to aircraft, each plot to a different aircraft, so as to maximise the product of the likelihoods. Determine the optimum assignment for the above data.

5   A market research team requires household data from each of 5 different towns. The team has $2\frac{1}{2}$ days at its disposal and intends to spend one half-day interviewing in each of the towns. The households scheduled for interview have been previously determined. From their previous experience the team estimate the probabilities of making a successful contact in each of the towns during the half days as follows.

Time period	A	B	C	D	E
Wednesday a.m.	0·67	0·62	0·52	0·40	0·63
Wednesday p.m.	0.90	0.70	0.65	0.87	0.83
Thursday a.m.	0·57	0·25	0·60	0·60	0·53
Thursday p.m.	0·40	0·52	0·45	0·43	0·50
Friday a.m.	0·63	0·60	0·40	0·36	0·67
No. of scheduled interviews	30	40	40	30	30

How should the team allocate the 5 half days to the 5 towns in order to maximise the expected number of successful interviews.

6 A small airline operates the following flights between three cities A, B, C. The flights operate daily seven days each week. The cost of stop-overs at the three airports is kT^2 where T is the stop-over time. How should the aeroplanes be assigned to the flights in order to minimise costs? You should assume that an aeroplane cannot take off again less than an hour after it landed since it has to be refuelled and essential checks must be carried out.

		Flight schedule		
Flight	From	Time of departure	To	Time of arrival
1	A	8.00	B	12.00
2	A	9.00	C	12.00
3	A	10.00	B	14.00
4	A	14.00	B	18.00
5	A	18.00	B	22.00
6	A	20.00	C	23.00
7	B	7.00	A	11.00
8	B	9.00	A	13.00
9	B	13.00	A	17.00
10	B	18.00	A	22.00
11	C	9.00	A	12.00
12	C	15.00	A	18.00

7 Use the Bradford method to solve the following assignment problem, where it is desired to minimise the total operative time in completing all the tasks.

			Time taken (hours)						
			Operative						
		1	2	3	4	5	6	7
	1	11	15	20	16	13	26	11
	2	12	13	22	14	16	29	13
	3	14	16	24	22	22	32	16
Task	4	14	12	20	19	20	31	15
	5	16	13	22	20	23	34	17
	6	13	15	18	14	26	29	18
	7	12	11	16	17	17	24	10

8 A company sells its product in 5 geographical regions. They estimate the sales potential of each region to be as follows:

Region	I	II	III	IV	V
Sales Potential ($)	80 000	60 000	50 000	40 000	20 000

Their 5 salesmen differ in ability. For each one it is reckoned that the proportion of potential sales that can be realised is as follows:

Salesman	A	B	C	D	E
Proportion	0.7	0.6	0.5	0.45	0.4

How should they assign the salesmen to the regions in order to maximise the total value of sales made?

9   Solve the problems of this chapter and these exercises using the transportation problem program. Do you notice a loss in efficiency compared with Mack's method?

10   Solve the problems of this chapter and these exercises using the Simplex Method program. Do you notice a loss in efficiency compared with Mack's method?

6
The Revised Simplex Method

6.1 The Revised Simplex Algorithm

The Simplex Method for the solution of a linear programming problem is not a very efficient computational procedure. The transformation from one canonical form to the next is carried out on all columns of coefficients. However, if the variable x_k never enters the basis during the procedure, the transformations on this column will have been to no purpose. Of course *a priori we do not know* which variables will become basic during the calculations.

A more efficient algorithm called the Revised Simplex Method was developed by Dantzig and Orchard-Hays in 1953. In many respects it follows the ideas underlying the Simplex Method but it has the advantage that it only calculates those quantities that are actually needed. It computes the simplex multipliers and inverse of the basis directly, whereas these were only apparent indirectly from the Simplex Method (see Chapter 3).

We consider in the first instance the case where the problem arises from '$\leqslant$' constraints so that a first basic feasible solution is provided by the slack variables. Thus our problem is:
find $x_i \geqslant 0$ $(i = 1, ..., n)$ such that

$$Ax \leqslant b, \tag{6.1}$$

which minimise

$$c^Tx = z, \tag{6.2}$$

where A is an $m \times n$ matrix of coefficients, b an $m \times 1$ column of non-negative values, and c^T a $1 \times n$ row of coefficients for the objective function.

Thus after the introduction of slack variables the problem in standard form becomes:

$$\left.\begin{array}{l} a_{11}x_1 + a_{12}x_2 + \cdots + a_{1n}x_n + x_{n+1} \qquad\qquad\qquad = b_1 \\ a_{21}x_1 + a_{22}x_2 + \cdots + a_{2n}x_n \qquad + x_{n+2} \qquad\quad = b_2 \\ \cdots\cdots\cdots\cdots\cdots\cdots\cdots\cdots\cdots\cdots\cdots\cdots\cdots\cdots\cdots\cdots\cdots \\ a_{m1}x_1 + a_{m2}x_2 + \cdots + a_{mn}x_n \qquad\qquad + x_{m+n} = b_m \\ c_1x_1 + \quad c_2x_2 + \cdots + c_nx_n \qquad\qquad\qquad\qquad = z \end{array}\right\} \tag{6.3}$$

(6.3) is of course the canonical form corresponding to the basic variables $x_{n+1}, ..., x_{n+m}$ which have values $b_1, b_2, ..., b_m$ respectively. The inverse of the basis is simply I_m, a unit $m \times m$ matrix. The corresponding simplex multipliers are $\pi_1 = 0$, $\pi_2 = 0, ..., \pi_m = 0$, since z as it stands does not involve the basic variables.

The Revised Simplex Method is an iterative procedure based on our knowing at the kth iteration:

(i) the basic variables and their values,
(ii) the inverse of the basis,
(iii) the simplex multipliers corresponding to the basis.

Thus we suppose that at iteration k, the basic variables are $x_1, x_2, ..., x_m$ (there is no loss of generality here) whose values are $b_1', b_2', ..., b_m'$. The inverse of the basis is the $m \times m$ matrix $\mathbf{B}^{-1}$ and the simplex multipliers have the values $\pi_1, \pi_2, ..., \pi_m$.

We proceed following the argument of the Simplex Method but we only calculate those quantities which are necessary, and we do not necessarily calculate them in the same way that they were calculated in the Simplex Method.

I   We first compute the coefficients of the non-basic variables in the canonical form for the objective function for this basis.

We use equation (3.7), so that

$$c_j' = c_j + \sum_{i=1}^{m} a_{ij}\pi_i = c_j + (\pi_1 \pi_2 ... \pi_m) \begin{pmatrix} a_{1j} \\ a_{2j} \\ a_{mj} \end{pmatrix}. \tag{6.4}$$

In equation (6.4) the π_i are the *current* simplex multipliers and the a_{ij} are *original* coefficients from equations (6.3).

The c_j' are calculated for each non-basic variable. They will of course be zero for basic variables. Then, following the Simplex Method, we find min $c_j' = c_s'$. If $c_s' < 0$
$$_j$$
variable x_s will enter the basis at the next iteration. If $c_s' \geqslant 0$ we have obtained the minimum for z.

II   To find the variable that x_s will replace we need the values of a_{is}' (just this column) in the current canonical form. These are obtained using the inverse of the basis. From equation (2.9)

$$a_s' = \mathbf{B}^{-1} a_s \tag{6.5}$$

i.e.

$$a_{is}' = \sum_{k=1}^{m} \beta_{ik} a_{ks} \tag{6.6}$$

where

$$\mathbf{B}^{-1} = \begin{pmatrix} \beta_{11} & \beta_{12} & \beta_{1m} \\ \beta_{21} & & \\ \beta_{m1} & & \beta_{mm} \end{pmatrix}. \tag{6.7}$$

The maximum value that can be given to x_s without violating the non-negativity conditions is then given by (refer to equation 2.14):

$$\max x_s = \min_{a_{is}' > 0} \frac{b_i'}{a_{is}'} = \frac{b_r'}{a_{rs}'}, \tag{6.8}$$

and this determines the row of the basic variable which leaves the basis.

III   We are now in a position to recalculate the new basis, its inverse, and the simplex multipliers.

(a) The old basis $(x_1, x_2, ..., x_r, ..., x_m)$ is replaced by $(x_1, x_2, ..., x_s, ..., x_m)$.

(b) The new values of the basic variables are

$$x_s = b_r^+ = \frac{b_r'}{a_{rs}'} \tag{6.9}$$

$$x_i = b_i^+ = b_i' - a_{is}' b_r^+ \quad (i \neq r). \tag{6.10}$$

(c) The inverse of the new basis can be obtained by recalling that the transformations of the Simplex Method, for the coefficients in the constraints, could be accomplished by premultiplying the current canonical form for the constraints by the matrix

rth column

$$E = \quad {}_{r\text{th row}} \begin{pmatrix} 1 & 0 & \dfrac{-a_{1s}'}{a_{rs}'} & 0 & 0 \\[2mm] 0 & 1 & \dfrac{-a_{2s}'}{a_{rs}'} & 0 & 0 \\[2mm] & & \cdots\cdots\cdots & & \\[2mm] 0 & 0 & \dfrac{1}{a_{rs}'} & 0 & 0 \\[2mm] & & \cdots\cdots\cdots & 1 & 0 \\[2mm] 0 & & \dfrac{-a_{ms}'}{a_{rs}'} & 0 & 1 \end{pmatrix}.$$

Thus   New Inverse $= E \times$ Old Inverse.   (6.11)

[See Question 15 of Exercises 3.]

Thus if   New Inverse $= (\beta_{ij}^+)$

$$\beta_{rj}^+ = \frac{\beta_{rj}}{a_{rs}'} \tag{6.12}$$

$$\beta_{ij}^+ = \beta_{ij} - \frac{a_{is}'}{a_{rs}'} \beta_{rj} = \beta_{ij} - a_{is}' \beta_{rj}^+ \quad (i \neq r). \tag{6.13}$$

Row r of the new inverse now refers to variable x_s. It should be noted that the transformations, both for the new values of the basic variables and the elements of the inverse matrix arise from Simplex type computations with

$$\begin{pmatrix} a_{1s}' \\ a_{2s}' \\ a_{rs}' \\ a_{ms}' \end{pmatrix}$$

as pivot column and a_{rs}' as pivot element.

(d) To complete the cycle we need the new simplex multipliers.
From equation (3.9) the old values were

$$\pi_i = - \sum_{k=1}^{m} \beta_{ki} c_k. \tag{6.14}$$

Thus

$$\pi_i^+ = - \sum_{\substack{k=1 \\ k \neq r}}^{m} \beta_{ki}^+ c_k - \beta_{ri}^+ c_s$$

since the basis now contains x_s instead of x_r.
Thus

$$\pi_i^+ = - \sum_{k=1}^{m} \left(\beta_{ki} - \frac{a_{ks}'}{a_{rs}'} \beta_{ri} \right) c_k - \frac{\beta_{ri}}{a_{rs}'} c_s$$

$$= - \sum_{k=1}^{m} \beta_{ki} c_k + \frac{\beta_{ri}}{a_{rs}'} \left(\sum_{k=1}^{m} a_{ks}' c_k - c_s \right)$$

since the term included by putting $k = r$ is zero.
Thus

$$\pi_i^+ = \pi_i + \frac{\beta_{ri}}{a_{rs}'} \left(\sum_{k=1}^{m} a_{ks}' c_k - c_s \right). \tag{6.15}$$

But the term in brackets is simply $-c_s'$. This can be seen by supposing that the constraints had been written in canonical form for the current basis. To obtain the objective function in the correct form we have to eliminate the basic variables. We can do this by multiplying the ith constraint by c_i and subtracting the result from the original form for z, for i ranging from 1 to m.
Thus

$$c_j' = c_j - \sum_{i=1}^{m} a_{ij}' c_i \quad (j = m+1, \dots, n).$$

[See also equation (2.10) and Section 3.2(ii).]
Thus

$$\pi_i^+ = \pi_i - \frac{\beta_{ri}}{a_{rs}'} c_s' = \pi_i - c_s' \beta_{ri}^+ \tag{6.16}$$

This transformation can also be viewed in the same light as those for the values and inverse, if we append the row of the π's to the inverse matrix and enhance the pivot column to

$$\begin{pmatrix} a_{1s}' \\ a_{2s}' \\ a_{ms}' \\ c_s' \end{pmatrix}.$$

In this way the calculations can be set out in table form as we pass from one iteration to the next.

Iteration	Basis	Value	Inverse of Basis				a'_{is}
	x_1	b'_1	β_{11}	β_{12}		β_{1m}	a'_{1s}
k	x_r	b'_r	β_{r1}	β_{r2}	β_{rs}	β_{rm}	a'_{rs}
	x_m	b'_m	β_{m1}	β_{m2}		β_{mm}	a'_{ms}
$-z$		$-z'_0$	π_1	π_2		π_m	c'_s
	x_1	$b'_1 - a'_{1s}b^+_r$	$\beta_{11} - a'_{1s}\beta^+_{r1}$			$\beta_{m1} - a'_{1s}\beta^+_{rm}$	
$k+1$	x_s	$b^+_r = \dfrac{b'_r}{a'_{rs}}$	$\beta^+_{r1} = \dfrac{\beta_{r1}}{a'_{rs}}$			$\beta^+_{rm} = \dfrac{\beta_{rm}}{a'_{rs}}$	
	x_m	$b'_m - a'_{ms}b^+_r$	$\beta_{m1} - a'_{ms}\beta^+_{r1}$			$\beta_{mm} - a'_{ms}\beta^+_{rm}$	
$-z$		$-z'_0 - c'_s b^+_r$	$\pi_1 - c'_s\beta^+_{r1}$			$\pi_m - c'_s\beta^+_{rm}$	

Example 1
Find non-negative x_1, x_2, x_3 which satisfy

$$x_1 + 2x_2 + 5x_3 \leqslant 36$$
$$2x_1 + 3x_2 + 3x_3 \leqslant 48$$
$$x_1 + x_2 + 2x_3 \leqslant 22$$

and minimise
$$-9x_1 - 10x_2 - 15x_3 = z.$$

If the problem is put in standard form the original form for the constraints and objective function is:

Constraint	Value	x_1	x_2	x_3	x_4	x_5	x_6
1	36	1	2	5	1	0	0
2	48	2	3	3	0	1	0
3	22	1	1	2	0	0	1
Objective	z	-9	-10	-15	0	0	0

This is preserved throughout the calculation. It would need to be stored in the computer in the event that the problem was solved by computer.

We next draw up a table in which we will record the c'_j at the various stages of the calculation. This appears below and is completed row by row at each iteration.

Iteration	c'_1	c'_2	c'_3	c'_4	c'_5	c'_6	
0	-9	-10	-15^*	0	0	0	x_3 to enter basis
1	-6^*	-4	0	3	0	0	x_1 to enter basis
2	0	-2^*	0	-1	0	10	x_2 to enter basis
3	0	0	0	$-\frac{1}{2}^*$	$\frac{3}{2}$	$\frac{13}{2}$	x_4 to enter basis
4	0	0	2	0	1	7	Optimum found

At iteration 0 our basis is x_4, x_5, x_6 with values 36, 48, 22. The inverse is simply I_3 and the three simplex multipliers are all 0. The calculation then proceeds according to the iterative scheme just described. The sequence of computations is to *write down* the basis, their values, the inverse and simplex multipliers for iteration 0. Then we calculate the c_j', find the variable s to enter the basis (asterisked), calculate the column a_{is}' (the pivot element is daggered) and then calculate the new basis, their values, the new inverse and the simplex multipliers. We then start the iterative process again. When all the c_j' are positive the minimum has been reached.

Iteration	Basis	Value	Inverse			a_{is}'	
0	x_4	36	1	0	0	5†	x_4 to leave basis
	x_5	48	0	1	0	3	
	x_6	22	0	0	1	2	
	$-z$	0	0	0	0	-15	
1	x_3	$\frac{36}{5}$	$\frac{1}{5}$	0	0	$\frac{1}{5}$	
	x_5	$\frac{132}{5}$	$-\frac{3}{5}$	1	0	$\frac{7}{5}$	
	x_6	$\frac{38}{5}$	$-\frac{2}{5}$	0	1	$\frac{3}{5}$†	x_6 to leave basis
	$-z$	108	3	0	0	-6	
2	x_3	$\frac{14}{3}$	$\frac{1}{3}$	0	$-\frac{1}{3}$	$\frac{1}{3}$	
	x_5	$\frac{26}{3}$	$\frac{1}{3}$	1	$-\frac{7}{3}$	$\frac{4}{3}$†	x_5 to leave basis
	x_1	$\frac{38}{3}$	$-\frac{2}{3}$	0	$\frac{5}{3}$	$\frac{1}{3}$	
	$-z$	184	-1	0	10	-2	
3	x_3	$\frac{5}{2}$	$\frac{1}{4}$	$-\frac{1}{4}$	$\frac{1}{4}$	$\frac{1}{4}$†	x_3 to leave basis
	x_2	$\frac{13}{2}$	$\frac{1}{4}$	$\frac{3}{4}$	$-\frac{7}{4}$	$\frac{1}{4}$	
	x_1	$\frac{21}{2}$	$-\frac{3}{4}$	$-\frac{1}{4}$	$\frac{9}{4}$	$-\frac{3}{4}$	
	$-z$	197	$-\frac{1}{2}$	$\frac{3}{2}$	$\frac{13}{2}$	$-\frac{1}{2}$	
4	x_4	10	1	-1	1		
	x_2	4	0	1	-2		
	x_1	18	0	-1	3		
	$-z$	202	0	1	7		

At iteration 4 we have the optimum solution in which $x_1 = 18$, $x_2 = 4$, $x_3 = 0$, $x_4 = 10$, $x_5 = 0$, $x_6 = 0$ and z has the minimum value -202.

The matrix

$$\begin{pmatrix} 1 & -1 & 1 \\ 0 & 1 & -2 \\ 0 & -1 & 3 \end{pmatrix}$$

is easily verified to be the inverse matrix of the columns in the original form corresponding to x_4, x_2, x_1 (in that order),

$$\begin{pmatrix} 1 & 2 & 1 \\ 0 & 3 & 2 \\ 0 & 1 & 1 \end{pmatrix}.$$

The simplex multipliers corresponding to the optimal basis are $\pi_1 = 0$, $\pi_2 = 1$ and $\pi_3 = 7$ as is easily verified.

It is perhaps worth pointing out that the transformation calculations have been carried out on a 4×4 matrix (of values, inverse elements and simplex multipliers). In the Simplex Method the calculations would have been carried out on a 4×7 matrix of values, and coefficients.

6.2   Initiating the Algorithm

In the general case when the constraints are a mixture of ' $\geqslant$ ', ' $=$ ' and ' $\leqslant$ ' types, a first basic feasible solution (and inverse matrix) is not immediately evident. We then have to adopt the strategy discussed in Section 2.3. We introduce artificial variables into the ' $\geqslant$ ' and ' $=$ ' constraints. These are in addition to the slack variables for the ' $\leqslant$ ' and ' $\geqslant$ ' constraints. The first basis for the modified constraints will consist of these artificial variables and the slacks from the ' $\leqslant$ ' constraints. The first inverse of the basis will be a unit matrix I_m when there are a *total* of m constraints.

The first problem (Phase I) is to minimise the artificial objective function w which is the sum of the artificial variables. We shall have a set of simplex multipliers σ_1, σ_2, ..., σ_m for w in addition to the simplex multipliers π_1, π_2, ..., π_m for z, the true objective function.

The coefficients of the non-basic variables in the canonical form for w will be denoted by the values d_j'.

In line with equation (6.4)

$$d_j' = d_j + \sum_{i=1}^{m} a_{ij}\sigma_i \tag{6.17}$$

where the original coefficients d_j are 0 for all the variables except the artificial variables. For these variables they are of course 1 since w is the sum of these variables.

Thus it becomes clear that in order to eliminate the basic variables from w at this stage the simplex multipliers σ_i must be -1 for those that correspond to the ' $\geqslant$ ' and ' $=$ ' constraints, and 0 for ' $\leqslant$ ' constraints. The π_i of course are all 0 as before. Whilst we are in Phase I the π_i and the σ_i are transformed in accordance with equation (6.16) using c_s' for the π_i and d_s' for the σ_i. c_j' and d_j' are computed from equations (6.4) and (6.17) respectively. In this way as soon as Phase I is completed and w has been reduced to 0 (otherwise the constraints will not have a feasible solution), we can continue without further ado with the minimisation of z.

Example 1

Find $x_1, x_2 \geqslant 0$ such that

$$3x_1 + 4x_2 \geqslant 19$$
$$3x_1 + x_2 \geqslant 7$$
$$x_1 + x_2 \leqslant 15$$

which minimise $10x_1 + 5x_2 = z.$

The problem in standard form with slack and artificial variables and artificial objective function is:

Constraint	Value	x_1	x_2	x_3	x_4	x_5	x_6	x_7
1	19	3	4	$-1.$	0	0	1	0
2	7	3	1	0	-1	0	0	1
3	15	1	1	0	0	1	0	0
Objective	z	10	5	0	0	0	0	0
Artificial	w	0	0	0	0	0	1	1

Iteration	c_1' or d_1'	c_2' or d_2'	c_3' or d_3'	c_4' or d_4'	c_5' or d_5'	c_6' or d_6'	c_7' or d_7'	
0	-6^*	-5	1	1	0	0	0	Phase 1. x_1 to enter basis.
1	0	-3^*	1	-1	0	0	2	Phase 1. x_2 to enter basis.
2	0	0	0	0	0	1	1	Phase 1 ends since all coeffs are $+$ve.
2	0	0	$\frac{5}{9}$	$\frac{25}{9}$	0			Phase 2 ends also since all coeffs. for z are positive.

Iteration	Basis	Value	Inverse			a_{is}'
0	x_6	19	1	0	0	3
	x_7	7	0	1	0	3†
	x_5	15	0	0	1	1
	$-z$	0	0	0	0	10
	$-w$	-26	-1	-1	0	-6
1	x_6	12	1	-1	0	3†
	x_1	$\frac{7}{3}$	0	$\frac{1}{3}$	0	$\frac{1}{3}$
	x_5	$\frac{38}{3}$	0	$-\frac{1}{3}$	1	$\frac{2}{3}$
	$-z$	$-\frac{70}{3}$	0	$-\frac{10}{3}$	0	$\frac{5}{3}$
	$-w$	-12	-1	1	0	-3
2	x_2	4	$\frac{1}{3}$	$-\frac{1}{3}$	0	
	x_1	1	$-\frac{1}{9}$	$\frac{4}{9}$	0	
	x_5	10	$-\frac{2}{9}$	$-\frac{1}{9}$	1	
	$-z$	-30	$-\frac{5}{9}$	$-\frac{25}{9}$	0	
	$-w$	0	0	0	0	

At iteration 2 w is reduced to zero. Using equation (6.4) and the appropriate simplex multipliers for z, viz. $-\frac{5}{9}$, $-\frac{25}{9}$, 0, it is seen that z is also optimised by this basic feasible solution, in which $x_1 = 1$, $x_2 = 4$ and the minimum for z is 30. Since $x_5 = 10$, the third constraint is satisfied as a strict inequality.

6.3   Degeneracy Revisited

It was pointed out in Section 2.5 that if a linear programming problem is degenerate then it is *possible* for the simplex iterations to *cycle* and return to a previous basis. Indeed our computer program did exhibit this phenomenon when it was used to solve Beale's problem. Fortunately we were able to overcome the difficulty through the simple device of re-ordering the constraints. However, a computer program needs to be robust and able to deal with all eventualities. We shall now take up the point again and resolve the difficulty.

In Section 2.5 it was shown that degeneracy arises because two or more vertices of the feasible region coincide. Of course the presence of degeneracy does not in itself make cycling inevitable, but for degeneracy to occur at a particular stage, a choice must have existed at the previous stage as to which variable was to leave the basis. Dantzig has shown that a suitable choice at this stage will prevent cycling and hence guarantee the convergence of the method. Dantzig arrived at his choice procedure by considering a perturbed problem. It was pointed out in Section 2.5, that by perturbing the problem, the coincidence of the vertices and hence the degeneracy could be removed.

Thus instead of considering the problem:
find $x_j \geqslant 0$ which satisfy

$$\sum_{j=1}^{n} a_{ij}x_j = b_i \quad (i = 1, \ldots, m) \tag{6.18}$$

and minimise

$$\sum_{j=1}^{n} c_j x_j = z,$$

we consider the problem (perturbed in a particular way):
find $x_j \geqslant 0$ which satisfy

$$\sum_{j=1}^{n} a_{ij}x_j = b_i + \varepsilon^i \quad (i = 1, \ldots, m) \tag{6.19}$$

and minimise

$$\sum_{j=1}^{n} c_j x_j = z.$$

Here ε is a non-negative arbitrary quantity. We shall normally think of it as being small. Indeed having solved the problem, our final optimal solution values will be functions of ε, and we shall let $\varepsilon \to 0$ so as to obtain a solution to the original problem. On the way to that solution, the presence of ε will have removed the degeneracy.

At the kth iteration we shall have a basis $(x_1, x_2, ..., x_m)$ say, and the inverse of the basis, the matrix with elements (β_{ij}). Thus from equation (2.8) the values of the basic variables for the perturbed problem will be given by

$$x_i(\varepsilon) = \sum_{k=1}^{m} \beta_{ik}(b_k + \varepsilon^k) = b_i' + \sum_{k=1}^{m} \beta_{ik}\varepsilon^k \qquad (6.20)$$

where b_i' are their values in the original problem.

Of course our choice of ε can then only be arbitrary to the extent that the basis remains feasible. If $b_i' > 0$, then because the elements β_{ik} are bounded, we can be sure that $x_i(\varepsilon) \geqslant 0$ by choosing ε sufficiently small. If $b_i' = 0$ then we must be sure that the first non-zero β_{ik} is positive, since $\varepsilon > 0$.

Now suppose that x_s is the variable to enter the basis. To find the variable to leave the basis we find the minimum value of the ratio

$$\frac{b_i' + \sum_{k=1}^{m} \beta_{ik}\varepsilon^k}{a_{is}'} \qquad (6.21)$$

for $a_{is}' > 0$.

Now if the original problem is degenerate it would mean that there could be a tie at this stage for the minimum of

$$\frac{b_i'}{a_{is}'} \qquad (6.22)$$

for $a_{is}' > 0$.

If the minimum of b_i'/a_{is}' for $a_{is}' > 0$ is unique and equal to b_r'/a_{rs}', then if ε is sufficiently small the minimum (6.21) will occur for the same value r.

The values of the basic variables at the next iteration will be, from equations (6.9) and (6.10):

$$x_s^+ = \frac{1}{a_{rs}'}\left(b_r' + \sum_{k=1}^{m} \beta_{rk}\varepsilon^k\right) \qquad (6.23)$$

and for $i \neq r$

$$x_i^+ = b_i' - \frac{a_{is}'}{a_{rs}'}b_r' + \sum_{k=1}^{m}\left(\beta_{ik} - \frac{a_{is}'}{a_{rs}'}\beta_{rk}\right)\varepsilon^k \qquad (6.24)$$

and, because of the way in which r was determined, the term independent of ε in equation (6.24) is strictly positive so that the solution is feasible.

If ties occur at (6.22) we have to consider terms to the first degree in ε in (6.21). Thus we search for the minimum of

$$\frac{b_i' + \beta_{i1}\varepsilon}{a_{is}'} \qquad (6.25)$$

among those values of i that tied at (6.22).

Thus we find r from

$$\min \frac{\beta_{i1}}{a'_{is}} = \frac{\beta_{r1}}{a'_{rs}}. \tag{6.26}$$

If this minimum is unique the values of the variables at the next iteration will be given by

$$x_s^+ = \frac{1}{a'_{rs}} \left(b'_r + \sum_{k=1}^{m} \beta_{rk} \varepsilon^k \right)$$

$$x_i^{++} = \sum_{k=1}^{m} \left(\beta_{ik} - \frac{a'_{is}}{a'_{rs}} \beta_{rk} \right) \varepsilon^k \tag{6.27}$$

for those i that tied at the first stage, and by equation (6.24) as before for other values of i.

Now because of equation (6.26) the coefficient of ε will be strictly positive in equation (6.27), thus ensuring that the basis remains feasible. If ties occur at (6.22), and again at (6.26), we have to consider the terms in ε^2 and determine r from

$$\min \frac{\beta_{i2}}{a'_{is}} = \frac{\beta_{r2}}{a'_{rs}} \tag{6.28}$$

where the minimum is taken over those i that tied before.

We continue this procedure, working across the columns of the inverse in the tied rows until the tie is resolved. This must occur at some stage, for otherwise it would imply that two rows of the inverse are identical, and that is impossible.

Thus we see that the procedure for finding the variable to leave the basis is to consider first the ratios at (6.22), and if ties occur to go to equation (6.26), and if ties still occur to go to equation (6.28) and so on across the columns of the inverse. Thus the situation can be resolved without *explicitly* introducing ε into the problem. It suffices that a value for ε satisfying all the conditions could be found. We do not have to find it.

The new value of z will be

$$z_0^+ = z_0' + c_s' x_s^+ \tag{6.29}$$

and because $c_s' < 0$ and $x_s^+ > 0$ (from equation (6.23))z is reduced and so cannot ever take on a previous value. Thus we can be sure that a previous *basis* cannot recur and so the method must terminate in at most $\binom{n}{m}$ steps. See Section 2.5.

Example 1

Find non-negative x_1, x_2, x_3 which satisfy

$$x_1 - 2x_2 \leqslant 2$$
$$2x_1 - x_2 \leqslant 4$$
$$x_1 + x_2 \leqslant 5$$

and minimise

$$-3x_1 + x_2 = z.$$

Constraint	Value	x_1	x_2	x_3	x_4	x_5
1	2	1	-2	1	0	0
2	4	2	-1	0	1	0
3	5	1	1	0	0	1
Objective	z	-3	1	0	0	0

Iteration	c_1'	c_2'	c_3'	c_4'	c_5'	
0	-3^*	1	0	0	0	x_1 enters basis.
1	0	$-\frac{1}{2}^*$	0	$\frac{3}{2}$	0	x_2 enters basis.
2	0	0	0	$\frac{4}{3}$	$\frac{1}{3}$	Optimum.

Iteration	Basis	Value	Inverse			a_{is}'	
0	x_3	2	1	0	0	1	
	x_4	4	0	1	0	$2\dagger$	x_4 leaves basis.
	x_5	5	0	0	1	1	
	$-z$	0	0	0	0	-3	
1	x_3	0	1	$-\frac{1}{2}$	0	$-\frac{3}{2}$	
	x_1	2	0	$\frac{1}{2}$	0	$-\frac{1}{2}$	
	x_5	3	0	$-\frac{1}{2}$	1	$\frac{3}{2}\dagger$	x_5 leaves basis.
	$-z$	6	0	$\frac{3}{2}$	0	$-\frac{1}{2}$	
2	x_3	3	1	-1	1		
	x_1	3	0	$\frac{1}{3}$	$\frac{1}{3}$		
	x_2	2	0	$-\frac{1}{3}$	$\frac{2}{3}$		
	$-z$	7	0	$\frac{4}{3}$	$\frac{1}{3}$		

The optimum for z is -7 when $x_1 = 3$ and $x_2 = 2$. At iteration 0, the tie between x_3 and x_4 as to which is to leave the basis is resolved by finding the smaller of

$$\frac{2 + \varepsilon + 0\varepsilon^2}{1} \quad \text{and} \quad \frac{4 + 0\varepsilon + \varepsilon^2}{2}.$$

Since ε is small it is clearly the latter. The tie in the 'values' $\frac{2}{1} = \frac{4}{2}$ is resolved in the first column of the inverse; $\frac{1}{1} > \frac{0}{2}$.

6.4 A Computer Program for the Revised Simplex Method

The computer PROGRAM *RevisedSimplex* which follows incorporates the ideas discussed in this chapter. It has features in common with the PROGRAMs of Chapter 2. In addition to the arrays a, b, c, and d which play the same roles as in the

PROGRAM *FullSimplex,* this new PROGRAM has arrays

$$bdash = \begin{pmatrix} b_1' \\ b_2' \\ b_m' \end{pmatrix} \qquad beta = \begin{pmatrix} \beta_{11} & \beta_{1m} \\ \beta_{m1} & \beta_{mm} \end{pmatrix} \qquad \begin{aligned} pi &= (\pi_1 \ \pi_m) \\ sigma &= (\sigma_1 \ \sigma_m). \end{aligned}$$

All values in arrays mentioned above are set up by either direct input in the PRO-CEDURE *inputdata* or by assignment in the PROCEDUREs *completetableau* and *setvalues.* The PROCEDURE *RevSimplex* implements the method described in Section 6.1. Comments in the program indicate where the computations required by equations (6.4) to (6.17) are performed. The rules of the previous section for breaking ties are incorporated into the latter part of the PROCEDURE *nextbasicvariable.* The reader is urged to compare the statements which begin IF NOT *unbounded* THEN in the PROCEDUREs *nextbasicvariable* of the PROGRAM *RevisedSimplex* which follows and the PROGRAM *FullSimplex* of Section 2.4.

As illustrations of the operation of the PROGRAM and the form of output we solve the examples given in Sections 6.1, 6.2 and 6.3. The output should be compared with the earlier computations. Finally we solve Beale's problem. Cycling is avoided. The reader might care to reverse the constraints just to check; there is still no cycling. The values in the CONST Section of the following PROGRAM are appropriate for the example in Section 6.1.

A few remarks on the computational advantages and disadvantages of the Revised Simplex Method and Simplex Method are perhaps in order. In the Revised Simplex Method the original matrix for the constraints is stored. In the Simplex Method it is overwritten as the iterations proceed. Thus, since we also need the inverse of the basis and simplex multipliers for the Revised Simplex Method, it does have the disadvantage of using more computer store.

Its great advantage lies in the reduction in the computation. The Simplex Method works on arrays a, b, c and d with a total of $mn + m + 2n$ elements. The Revised Simplex Method works on arrays *beta, bdash, pi* and *sigma* with a total of $m^2 + 3m$ elements. Thus we see that the computation in a linear programming problem is related directly to the number of constraints rather than the number of variables. In general the latter will be much larger.

The Revised Simplex Method computes directly the inverse of the basis and the simplex multipliers. We do not have to pick these quantities out from a final tableau. This can help in any subsequent sensitivity analysis.

```
PROGRAM RevisedSimplex (input,output);
CONST
    nvar=3; m=3;            { No. of variables and constraints }    {**}
    ncols=6;                { Maximum no. of columns in tableau }   {**}
    fwt=7;  dpt=2;          { Output format constants for tableau values }  {**}
    fwi=1;                  { Output format constant for indices }  {**}
    largevalue = 1.0E20;  smallvalue=1.0E-10;                       {**}
```

```
TYPE   mrange = 1..m;   ncolsrange = 1..ncols;
  matrix = ARRAY [mrange,ncolsrange] OF real;
  msquare = ARRAY [mrange,mrange] OF real;
  column = ARRAY [mrange] OF real;
  baseindex = ARRAY [mrange] OF integer;
  row = ARRAY [ncolsrange] OF real;
  rowboolean = ARRAY [ncolsrange] OF boolean;
  phase = (PhaseI, PhaseII);


VAR
  a : matrix;   { Matrix A in standard form of problem, see (2.3) }
  b : column;   { Vector b in standard form of problem, see (2.3) }
  c : row;      { Coefficients of objective function, see (2.1) }
  d : row;      { Coefficients of artificial objective function }
  basic : baseindex;       { Basic variables at each stage }
  nonbasic : rowboolean;   { Status indicators for variables }
  w0, z0 : real;           { Values of objective functions }
  beta : msquare;          { Inverse of matrix B, see (2.9) & (6.5) }
  bdash : column;          { Values of basic variables }
  pi, sigma : column;      { Simplex multipliers, see (6.4) & (6.17) }
  it : integer;            { Iteration counter }
  solution, OK : boolean;  { Iteration process terminators }
  r, s : integer;          { Row and column of pivot element }
  GC, EC, LC : integer;
  n1, n2, GCplusLC, GCplusEC, blanksleft, blanksright : integer;
  printon : boolean;   i : integer;   slack : row;


PROCEDURE inputdata;
VAR i, j, k : integer;
BEGIN  read(k); printon:=k>0; read(GC,EC,LC);
  FOR i:=1 TO m DO
  BEGIN FOR j:=1 TO nvar DO read(a[i,j]);   read(b[i]) END;
  FOR j:=1 TO nvar DO read(c[j])
END; { inputdata }


PROCEDURE initialise;
VAR i, j, k : integer;
BEGIN  it:=0; z0:=0.0; OK:=true; GCplusLC:=GC+LC; GCplusEC:=GC+EC;
  n1 := nvar + GCplusLC + GCplusEC;   n2 := nvar + GCplusLC;
  FOR j:= nvar+1 TO n1 DO
  BEGIN FOR i:=1 TO m DO a[i,j]:=0.0; c[j]:=0.0 END;
  FOR i:=1 TO m DO FOR j:=1 TO m DO beta[i,j]:=0.0;
  FOR j:=1 TO m DO pi[j]:=0.0;   k:=m*fwt;   IF k<18 THEN k:=18;
  blanksright := (k-16) DIV 2;   blanksleft := k-16-blanksright
END; { initialise }


PROCEDURE completetableau;
VAR  i, j : integer;   sum : real;
```

```
BEGIN  FOR i:=1 TO GC DO a[i,nvar+i] := -1.0;
  FOR i:=1 TO LC DO a[GCplusEC+i,nvar+GC+i] := 1.0;
  FOR i:=1 TO GCplusEC DO a[i,nvar+GCplusLC+i] := 1.0;
  { Compute initial base and w0 }
  FOR j:=1 TO GCplusEC DO basic[j] := nvar + GCplusLC + j;
  FOR j:=1 TO LC DO basic[GCplusEC+j] := nvar + GC + j;
  FOR j:=1 TO nl DO nonbasic[j]:=true;
  FOR i:=1 TO m DO nonbasic[basic[i]]:=false;
  sum:=0.0;  FOR i:=1 TO GCplusEC DO sum := sum + b[i];  w0:=-sum
END; { completetableau }


PROCEDURE setvalues;
{ Set up values for Revised Simplex Method calculations }
VAR  i, j : integer;
BEGIN
  { Initialise artificial simplex multipliers }
  FOR j:=1 TO GCplusEC DO sigma[j]:=-1.0;
  FOR j:=GCplusEC+1 TO GCplusEC+LC DO sigma[j]:=0.0;
  { Set leading diagonal of inverse matrix }
  FOR i:=1 TO m DO beta[i,i]:=1.0;
  { Initialise d-values }
  FOR j:=1 TO n2 DO d[j]:=0.0;
  FOR j:=n2+1 TO nl DO d[j]:=1.0;
  { Initialise basic variable values }
  FOR j:=1 TO m DO bdash[j]:=b[j]
END; { setvalues }


PROCEDURE outputtableau (p:phase);
VAR  i, j, n : integer;
BEGIN  n:=nl;
  writeln; writeln(´   INITIAL TABLEAU´);
  write(´   BASE VAR.   ´, ´ ´:fwt-5, ´VALUE´);
  FOR j:=1 TO n DO write(´ ´:fwt-fwi-1, ´X´, j:fwi); writeln;
  FOR i:=1 TO m DO
  BEGIN
    write(´ ´:8-fwi, ´X´, basic[i]:fwi, ´ ´:7, b[i]:fwt:dpt);
    FOR j:=1 TO n DO write(a[i,j]:fwt:dpt); writeln
  END;
  write(´   OBJECTIVE´, ´ ´:fwt, ´Z ´);
  FOR j:=1 TO n DO write(c[j]:fwt:dpt);  writeln;
  IF p=PhaseI THEN
  BEGIN  write(´   ARTIFICIAL´, ´ ´:fwt-1, ´W ´);
    FOR j:=1 TO n DO write(d[j]:fwt:dpt);  writeln
  END
END; { outputtableau }


PROCEDURE RevSimplex (p:phase);
VAR  n : integer;  unbounded : boolean;  v : column;  corddash : row;
     csdash : real;  ch : char;


  PROCEDURE initialise;
  BEGIN  solution:=false;  unbounded:=false END; { initialise }
```

```
PROCEDURE printdetails;
VAR  i, j : integer;  finished : boolean;
BEGIN  finished := (p=phaseII) AND solution;
  write('       BASE VAR. ' , ' ':fwt-5, ' VALUE',
          ' ':blanksleft, 'INVERSE OF BASIS');
  IF NOT finished THEN write(' ':blanksright, ' ':fwt-5, 'A''[I]');
  writeln;
  FOR i:=1 TO m DO
  BEGIN
    write(' ':10-fwi, 'X', basic[i]:fwi, ' ':7, bdash[i]:fwt:dpt);
    FOR j:=1 TO m DO write(beta[i,j]:fwt:dpt);
    IF NOT finished THEN write(v[i]:fwt:dpt);  writeln;
  END;
  write('    OBJ SIM MULT ', z0:fwt:dpt);
  FOR j:=1 TO m DO write(pi[j]:fwt:dpt);
  IF p=PhaseI THEN writeln(csdash:fwt:dpt)
  ELSE IF NOT finished THEN writeln(corddash[s]:fwt:dpt);
  IF p=PhaseI THEN
  BEGIN  write('    ART SIM MULT ', w0:fwt:dpt);
    FOR j:=1 TO m DO write(sigma[j]:fwt:dpt);
    writeln(corddash[s]:fwt:dpt);
  END
END; { printdetails }

PROCEDURE nextbasicvariable (VAR r,s:integer; x:row; y:column);
VAR  i, j, k : integer;  min : real;
     sum, quotient, diff : real;  tiebroken : boolean;
BEGIN  min:=largevalue;  { Find the variable, s, }
  FOR j:=1 TO n DO        { to enter the basis.  }
    BEGIN  sum :=0.0;
      FOR i:=1 TO m DO sum := sum + y[i]*a[i,j];
      corddash[j] := x[j] + sum;
      IF nonbasic[j] THEN
        IF corddash[j]<min THEN BEGIN min:=corddash[j]; s:=j END
    END;
    IF p=PhaseI THEN
    BEGIN  sum:=0.0;
      FOR i:=1 TO m DO sum := sum + pi[i]*a[i,s];
      csdash := sum + c[s]
    END;
  solution := corddash[s] > -smallvalue;
  IF NOT (solution AND (p=phaseI)) THEN
  BEGIN  IF p=PhaseI THEN ch:='D' ELSE ch:= 'C';
    IF printon THEN
    BEGIN  writeln;  writeln('    ITERATION ', it:2);
      write('      '); FOR j:=1 TO n DO write(' ':fwt-fwi-1, ch, j:fwi);
      writeln; write('     ');
      FOR j:=1 TO n DO write(corddash[j]:fwt:dpt); writeln
    END
  END;
  IF NOT solution THEN
  BEGIN
    IF printon THEN writeln('    X', s:fwi, ' ENTERS BASIS');
    FOR i:=1 TO m DO  { Calculate the a'[i,s], see (6.6) }
```

```
      BEGIN  sum:=0.0;  { and store this column in v.      }
        FOR j:=1 TO m DO sum := sum + beta[i,j]*a[j,s] ;  v[i]:=sum;
      END;
      unbounded:=true;  i:=1;         { Check that at least one value }
      WHILE unbounded AND (i<=m) DO  { in column s is positive.     }
      BEGIN  unbounded := v[i] < smallvalue;  i:=i+1 END;
      IF NOT unbounded THEN
      BEGIN  min:=largevalue;  { Find the variable, basic[r], }
        FOR i:=1 TO m DO         { to leave the basis.           }
          IF v[i] > smallvalue THEN
          BEGIN  k:=0;
            REPEAT
              IF k=0 THEN quotient := bdash[i]/v[i]
                     ELSE quotient := beta[i,k]/v[i];
              diff := quotient - min;
              IF diff < -smallvalue THEN
              BEGIN  min:=bdash[i]/v[i]; r:=i; tiebroken:=true  END
              ELSE
                IF abs(diff) < smallvalue THEN  { Tiebreak required }
                BEGIN  k:=k+1;                  { see section 6.3.  }
                  min := beta[r,k]/v[r];  tiebroken:=false
                END
                ELSE tiebroken:=true
            UNTIL tiebroken
          END;
        IF printon THEN
        BEGIN  printdetails;
          writeln('    X', basic[r]:fwi, ' FROM CONSTRAINT ', r:fwi,
                  ' LEAVES THE BASIS')
        END;
        nonbasic[basic[r]]:=true; nonbasic[s]:=false; basic[r]:=s;
      END
    END
END; { nextbasicvariable }



PROCEDURE transformvalues (r,s:integer; VAR y:column; VAR y0:real);
{ Calculate the new values of the basic variables, the new inverse }
{ matrix and the new simplex multipliers, see Section 6.1 part III }
VAR  i, j : integer;  pivot : real;
BEGIN  pivot:=v[r];  bdash[r] := bdash[r]/pivot;  {(6.9)}
  FOR j:=1 TO m DO beta[r,j] := beta[r,j]/pivot;  {(6.12)}
  FOR i:=1 TO m DO
  BEGIN
    IF i<>r THEN
    BEGIN  bdash[i] := bdash[i] - v[i]*bdash[r];  {(6.10)}
      FOR j:=1 TO m DO
        beta[i,j] := beta[i,j] - v[i]*beta[r,j]  {(6.13)}
    END;
    y[i] := y[i] - corddash[s]*beta[r,i]  {(6.16)}
  END;
  y0 := y0 - corddash[s]*bdash[r];
  IF p=PhaseI THEN
```

```
    BEGIN
      FOR i:=1 TO m DO pi[i] := pi[i] - csdash*beta[r,i];
      z0 := z0 - csdash*bdash[r]
    END
  END; { transformvalues }



BEGIN   { RevSimplex }
  initialise;
  IF p=PhaseI THEN n:=n1 ELSE n:=n2;
  REPEAT
    CASE p OF
      PhaseI  : BEGIN
                    nextbasicvariable(r,s,d,sigma);
                    IF NOT (solution OR unbounded) THEN
                    BEGIN transformvalues(r,s,sigma,w0); it:=it+1 END
                END;
      PhaseII : BEGIN
                    nextbasicvariable(r,s,c,pi);
                    IF NOT (solution OR unbounded) THEN
                    BEGIN transformvalues(r,s,pi,z0); it:=it+1 END
                END
    END
  UNTIL solution OR unbounded;
  IF unbounded THEN writeln('   UNBOUNDED')
  ELSE IF (p=PhaseII) AND printon THEN printdetails
END; { RevSimplex }



BEGIN   { Main Program }
  writeln; writeln('   REVISED SIMPLEX METHOD'); writeln;
  inputdata; initialise; completetableau; setvalues;
  IF GCplusEC=0 THEN
  BEGIN writeln('   THERE IS NO PHASE I');
    outputtableau(PhaseII); writeln
  END
  ELSE  { Perform Phase I }
  BEGIN writeln('   PHASE I');
    outputtableau(PhaseI);
    RevSimplex(PhaseI);  writeln;
    IF (abs(w0)>smallvalue) OR (NOT solution) THEN
    BEGIN  OK:=false;  writeln('   PHASE I NOT COMPLETED');
           writeln('   SUM OF ARTIFICIALS ', w0:fwt:dpt)
    END
    ELSE
    BEGIN writeln('   PHASE I SUCCESSFUL');
      writeln; writeln('   PHASEII');
    END
  END;
  IF OK THEN   { Perform Phase II }
  BEGIN RevSimplex(PhaseII);  writeln;
    IF NOT solution THEN writeln('   PHASE II NOT COMPLETED')
    ELSE
```

```
    BEGIN  { Output final details }
      writeln; writeln('    FINAL SOLUTION'); writeln;
      writeln('    MINIMUM OF Z = ', -z0:fwt:dpt); writeln;
      write('    CONSTRAINT    BASIS      VALUE');
      writeln('    STATE      SLACK         PI');
      FOR i:=1 TO m DO slack[basic[i]] := bdash[i];
      FOR i:=1 TO m DO { For each constraint }
      BEGIN
        write(i:10, basic[i]:10, ' ':12-fwt, bdash[i]:fwt:dpt, ' ':5);
        IF (i<=GC) OR (i>GCplusEC) THEN
          IF nonbasic[nvar+i] THEN write('BINDING', 0.0:10:dpt)
          ELSE  write('SLACK', ' ':12-fwt, slack[nvar+i]:fwt:dpt)
        ELSE  write('EQUATION      NONE');
        writeln(' ':12-fwt, pi[i]:fwt:dpt)
      END
    END
  END
END. { RevisedSimplex }
```

```
REVISED SIMPLEX METHOD

THERE IS NO PHASE I

INITIAL TABLEAU
BASE VAR.      VALUE      X1      X2      X3      X4      X5      X6
   X4          36.00    1.00    2.00    5.00    1.00    0.00    0.00
   X5          48.00    2.00    3.00    3.00    0.00    1.00    0.00
   X6          22.00    1.00    1.00    2.00    0.00    0.00    1.00
OBJECTIVE        Z      -9.00  -10.00  -15.00    0.00    0.00    0.00


ITERATION  0
      C1      C2      C3      C4      C5      C6
    -9.00  -10.00  -15.00    0.00    0.00    0.00
X3 ENTERS BASIS
   BASE VAR.     VALUE    INVERSE OF BASIS       A'[I]
      X4         36.00    1.00    0.00    0.00    5.00
      X5         48.00    0.00    1.00    0.00    3.00
      X6         22.00    0.00    0.00    1.00    2.00
OBJ SIM MULT     0.00    0.00    0.00    0.00  -15.00
X4 FROM CONSTRAINT 1 LEAVES THE BASIS

ITERATION  1
      C1      C2      C3      C4      C5      C6
    -6.00   -4.00    0.00    3.00    0.00    0.00
X1 ENTERS BASIS
   BASE VAR.     VALUE    INVERSE OF BASIS       A'[I]
      X3          7.20    0.20    0.00    0.00    0.20
      X5         26.40   -0.60    1.00    0.00    1.40
      X6          7.60   -0.40    0.00    1.00    0.60
OBJ SIM MULT   108.00    3.00    0.00    0.00   -6.00
X6 FROM CONSTRAINT 3 LEAVES THE BASIS
```

ITERATION 2

C1	C2	C3	C4	C5	C6
0.00	-2.00	0.00	-1.00	0.00	10.00

X2 ENTERS BASIS

BASE VAR.	VALUE	INVERSE OF BASIS			A´[I]
X3	4.67	0.33	0.00	-0.33	0.33
X5	8.67	0.33	1.00	-2.33	1.33
X1	12.67	-0.67	0.00	1.67	0.33
OBJ SIM MULT	184.00	-1.00	0.00	10.00	-2.00

X5 FROM CONSTRAINT 2 LEAVES THE BASIS

ITERATION 3

C1	C2	C3	C4	C5	C6
0.00	0.00	0.00	-0.50	1.50	6.50

X4 ENTERS BASIS

BASE VAR.	VALUE	INVERSE OF BASIS			A´[I]
X3	2.50	0.25	-0.25	0.25	0.25
X2	6.50	0.25	0.75	-1.75	0.25
X1	10.50	-0.75	-0.25	2.25	-0.75
OBJ SIM MULT	197.00	-0.50	1.50	6.50	-0.50

X3 FROM CONSTRAINT 1 LEAVES THE BASIS

ITERATION 4

C1	C2	C3	C4	C5	C6
0.00	0.00	2.00	0.00	1.00	7.00

BASE VAR.	VALUE	INVERSE OF BASIS		
X4	10.00	1.00	-1.00	1.00
X2	4.00	0.00	1.00	-2.00
X1	18.00	0.00	-1.00	3.00
OBJ SIM MULT	202.00	0.00	1.00	7.00

FINAL SOLUTION

MINIMUM OF Z = -202.00

CONSTRAINT	BASIS	VALUE	STATE	SLACK	PI
1	4	10.00	SLACK	10.00	0.00
2	2	4.00	BINDING	0.00	1.00
3	1	18.00	BINDING	0.00	7.00

REVISED SIMPLEX METHOD

PHASE I

INITIAL TABLEAU

BASE VAR.	VALUE	X1	X2	X3	X4	X5	X6	X7
X6	19.00	3.00	4.00	-1.00	0.00	0.00	1.00	0.00
X7	7.00	3.00	1.00	0.00	-1.00	0.00	0.00	1.00
X5	15.00	1.00	1.00	0.00	0.00	1.00	0.00	0.00
OBJECTIVE	Z	10.00	5.00	0.00	0.00	0.00	0.00	0.00
ARTIFICIAL	W	0.00	0.00	0.00	0.00	0.00	1.00	1.00

```
ITERATION   0
      D1        D2        D3        D4        D5        D6        D7
    -6.00    -5.00      1.00      1.00      0.00      0.00      0.00
X1 ENTERS BASIS
   BASE VAR.       VALUE     INVERSE OF BASIS      A´[I]
      X6          19.00     1.00     0.00     0.00     3.00
      X7           7.00     0.00     1.00     0.00     3.00
      X5          15.00     0.00     0.00     1.00     1.00
OBJ SIM MULT       0.00     0.00     0.00     0.00    10.00
ART SIM MULT     -26.00    -1.00    -1.00     0.00    -6.00
X7 FROM CONSTRAINT 2 LEAVES THE BASIS


ITERATION   1
      D1        D2        D3        D4        D5        D6        D7
     0.00    -3.00      1.00     -1.00      0.00      0.00      2.00
X2 ENTERS BASIS
   BASE VAR.       VALUE     INVERSE OF BASIS      A´[I]
      X6          12.00     1.00    -1.00     0.00     3.00
      X1           2.33     0.00     0.33     0.00     0.33
      X5          12.67     0.00    -0.33     1.00     0.67
OBJ SIM MULT     -23.33     0.00    -3.33     0.00     1.67
ART SIM MULT     -12.00    -1.00     1.00     0.00    -3.00
X6 FROM CONSTRAINT 1 LEAVES THE BASIS


PHASE I SUCCESSFUL


PHASEII


ITERATION   2
      C1        C2        C3        C4        C5
     0.00      0.00      0.56      2.78      0.00
   BASE VAR.       VALUE     INVERSE OF BASIS
      X2           4.00     0.33    -0.33     0.00
      X1           1.00    -0.11     0.44     0.00
      X5          10.00    -0.22    -0.11     1.00
OBJ SIM MULT     -30.00    -0.56    -2.78     0.00


FINAL SOLUTION


MINIMUM OF Z =    30.00


CONSTRAINT    BASIS      VALUE      STATE       SLACK          PI
    1           2        4.00      BINDING      0.00        -0.56
    2           1        1.00      BINDING      0.00        -2.78
    3           5       10.00      SLACK       10.00         0.00
```

REVISED SIMPLEX METHOD

THERE IS NO PHASE I

INITIAL TABLEAU

BASE VAR.	VALUE	X1	X2	X3	X4	X5
X3	2.00	1.00	-2.00	1.00	0.00	0.00
X4	4.00	2.00	-1.00	0.00	1.00	0.00
X5	5.00	1.00	1.00	0.00	0.00	1.00
OBJECTIVE	Z	-3.00	1.00	0.00	0.00	0.00

ITERATION 0

C1	C2	C3	C4	C5
-3.00	1.00	0.00	0.00	0.00

X1 ENTERS BASIS

BASE VAR.	VALUE	INVERSE OF BASIS			A´[I]
X3	2.00	1.00	0.00	0.00	1.00
X4	4.00	0.00	1.00	0.00	2.00
X5	5.00	0.00	0.00	1.00	1.00
OBJ SIM MULT	0.00	0.00	0.00	0.00	-3.00

X4 FROM CONSTRAINT 2 LEAVES THE BASIS

ITERATION 1

C1	C2	C3	C4	C5
0.00	-0.50	0.00	1.50	0.00

X2 ENTERS BASIS

BASE VAR.	VALUE	INVERSE OF BASIS			A´[I]
X3	0.00	1.00	-0.50	0.00	-1.50
X1	2.00	0.00	0.50	0.00	-0.50
X5	3.00	0.00	-0.50	1.00	1.50
OBJ SIM MULT	6.00	0.00	1.50	0.00	-0.50

X5 FROM CONSTRAINT 3 LEAVES THE BASIS

ITERATION 2

C1	C2	C3	C4	C5
0.00	0.00	0.00	1.33	0.33

BASE VAR.	VALUE	INVERSE OF BASIS		
X3	3.00	1.00	-1.00	1.00
X1	3.00	0.00	0.33	0.33
X2	2.00	0.00	-0.33	0.67
OBJ SIM MULT	7.00	0.00	1.33	0.33

FINAL SOLUTION

MINIMUM OF Z =   -7.00

CONSTRAINT	BASIS	VALUE	STATE	SLACK	PI
1	3	3.00	SLACK	3.00	0.00
2	1	3.00	BINDING	0.00	1.33
3	2	2.00	BINDING	0.00	0.33

REVISED SIMPLEX METHOD

THERE IS NO PHASE I

INITIAL TABLEAU

BASE VAR.	VALUE	X1	X2	X3	X4	X5	X6	X7
X5	0.00	0.25	-8.00	-1.00	9.00	1.00	0.00	0.00
X6	0.00	0.50	-12.00	-0.50	3.00	0.00	1.00	0.00
X7	1.00	0.00	0.00	1.00	0.00	0.00	0.00	1.00
OBJECTIVE	Z	-0.75	20.00	-0.50	6.00	0.00	0.00	0.00

ITERATION 0

C1	C2	C3	C4	C5	C6	C7
-0.75	20.00	-0.50	6.00	0.00	0.00	0.00

X1 ENTERS BASIS

BASE VAR.	VALUE	INVERSE OF BASIS			A'[I]
X5	0.00	1.00	0.00	0.00	0.25
X6	0.00	0.00	1.00	0.00	0.50
X7	1.00	0.00	0.00	1.00	0.00
OBJ SIM MULT	0.00	0.00	0.00	0.00	-0.75

X6 FROM CONSTRAINT 2 LEAVES THE BASIS

ITERATION 1

C1	C2	C3	C4	C5	C6	C7
0.00	2.00	-1.25	10.50	0.00	1.50	0.00

X3 ENTERS BASIS

BASE VAR.	VALUE	INVERSE OF BASIS			A'[I]
X5	0.00	1.00	-0.50	0.00	-0.75
X1	0.00	0.00	2.00	0.00	-1.00
X7	1.00	0.00	0.00	1.00	1.00
OBJ SIM MULT	0.00	0.00	1.50	0.00	-1.25

X7 FROM CONSTRAINT 3 LEAVES THE BASIS

ITERATION 2

C1	C2	C3	C4	C5	C6	C7
0.00	2.00	0.00	10.50	0.00	1.50	1.25

BASE VAR.	VALUE	INVERSE OF BASIS		
X5	0.75	1.00	-0.50	0.75
X1	1.00	0.00	2.00	1.00
X3	1.00	0.00	0.00	1.00
OBJ SIM MULT	1.25	0.00	1.50	1.25

FINAL SOLUTION

MINIMUM OF Z =   -1.25

CONSTRAINT	BASIS	VALUE	STATE	SLACK	PI
1	5	0.75	SLACK	0.75	0.00
2	1	1.00	BINDING	0.00	1.50
3	3	1.00	BINDING	0.00	1.25

Exercises 6

1 Use the Revised Simplex Method to solve the problem:
find $x_1, x_2 \geqslant 0$ such that

$$3x_1 + 4x_2 \leqslant 1700$$
$$2x_1 + 5x_2 \leqslant 1600$$

and minimise $\quad -2x_1 - 4x_2 = z.$

2 Solve, using the Revised Simplex Method:
find $x_i \geqslant 0, i = 1, 2,$
such that
$$x_1 \qquad \geqslant 10$$
$$x_2 \geqslant 5$$
$$x_1 + x_2 \leqslant 20$$
$$-x_1 + 4x_2 \leqslant 20$$
which maximise $\quad 3x_1 + 4x_2 = z.$

3 Solve, using the Revised Simplex Method:
find $x_1, x_2 \geqslant 0$

such that
$$x_1 + 3x_2 \geqslant 8$$
$$3x_1 + 4x_2 \geqslant 19$$
$$3x_1 + x_2 \geqslant 7$$
which minimise $\quad 50x_1 + 25x_2 = z.$

4 Solve, using the Revised Simplex Method:
find, $y_1, y_2, y_3 \geqslant 0$

$$y_1 + 3y_2 + 3y_3 \leqslant 50$$
$$3y_1 + 4y_2 + y_3 \leqslant 25$$
which maximise $\quad 8y_1 + 19y_2 + 7y_3 = w.$

5 Use the Revised Simplex Method to solve Beale's problem:
find $x_1, x_2, x_3, x_4 \geqslant 0$

such that
$$\tfrac{1}{2}x_1 - 12x_2 - \tfrac{1}{2}x_3 + 3x_4 \leqslant 0$$
$$\tfrac{1}{4}x_1 - 8x_2 - x_3 + 9x_4 \leqslant 0$$
$$x_3 \leqslant 1$$
which minimise $\quad -\tfrac{3}{4}x_1 + 20x_2 - \tfrac{1}{2}x_3 + 6x_4 = z.$

Verify that the rules for preventing cycling work whichever way the constraints are ordered.

6 A company produces a variety of items of office furniture. It can produce three types of desk. The amount of labour required for each, in the three production processes, is given in the table on the next page.

	Labour (man hours) to produce one desk of type		
	I	II	III
Component production	2	3	2
Component assembly	1	2	3
Polishing and inspection	1	1	2

The available labour each week for component production is 360 man hours, for assembly, 240 man hours and for polishing and inspection 180 man hours. The sales market is buoyant but despatch and storage facilities restrict total production to at most 170 desks per week.

The profit on each desk of type I, II, III is $15, $22 and $19 respectively. Show that the problem of obtaining the optimum production schedule can be put in the form:

find $x_i \geqslant 0$ such that

$$
\begin{aligned}
2x_1 + 3x_2 + x_3 + x_4 &= 360 \\
x_1 + 2x_2 + 3x_3 + x_5 &= 240 \\
x_1 + x_2 + 2x_3 + x_6 &= 180 \\
x_1 + x_2 + x_3 + x_7 &= 170
\end{aligned}
$$

and $w = -15x_1 - 22x_2 - 19x_3$ is a minimum.

Explain the physical significance of the seven variables in this model and, using the Revised Simplex Method, carry out the calculations necessary to obtain the solution.

The computer solution of the problem was given in the form:

Constraint	Basis	Values	Inverse of Basis				Simplex Multiplier
1	x_1	100	$\frac{1}{3}$	$-\frac{4}{3}$	$\frac{5}{3}$	0	6
2	x_2	40	$\frac{1}{3}$	$\frac{2}{3}$	$-\frac{4}{3}$	0	1
3	x_3	20	$-\frac{1}{3}$	$\frac{1}{3}$	$\frac{1}{3}$	0	2
4	x_7	10	$-\frac{1}{3}$	$\frac{1}{3}$	$-\frac{2}{3}$	1	0
Objective	w	-2760					

Use this result to construct the complete simplex tableau corresponding to the optimal solution.

The extra cost of overtime in each of the manufacturing stages is $4 per man hour. In which of the production processes is it worthwhile to use overtime? During a certain period it is necessary, in order to meet a particular order from a valued customer, to increase production of type III desks to at least 30 per week. How does this modify the optimal production schedule?

7 A company produces three types of drilling machine D_1, D_2, D_3 with profits of $10, $10 and $30, respectively, on models of each type. The numbers which can be manufactured in one week are limited by the supplies of fittings F_1, F_2 and F_3 where D_1 requires 1 of F_1, 4 of F_2 and 2 of F_3, D_2 requires 2 of F_1, 3 of F_2 and 3 of F_3 and D_3 requires 10 of F_1, 10 of F_2 and 8 of F_3. In any week the total availabilities of F_1, F_2, F_3 are 650, 850 and 650 respectively. Determine the maximum profit which can be achieved in any week and show that this maximum is achieved by making equal numbers of D_1, D_2 and D_3. The company applies to the Price Commission for permission to increase its prices to levels which give a 20% increase in the profits on all models. After investigation the Commission allows the increases in the prices of D_1 and D_2 but insists on a cut in the price of D_3 which has the effect of reducing the profit on D_3 by 10%. Show that an optimal production schedule based on these profit figures would lead to a 20% increase in the overall profit.

The management has an agreement which guarantees employment to 300 of its workers. If the labour requirements are such that one worker can produce one of D_1 per week or one of D_2 per week and five workers are needed to produce one D_3 per week, what effect does this agreement have on the new solution? Show that the maximum increase in profits which is realisable is 11% (over the original profit).

8 A simple model of the agricultural industry, for the export market, of the Narva Islands has been proposed along the following lines. There are three main crops which are suited to the climatic conditions and these can be grown on either of two types of arable land. At present 14×10^5 acres of type I and 12×10^5 acres of type II land are available for cultivation. The land types give different yields for the different crops and it is estimated that the net revenue from 1 acre of crop i grown on land of type j will be R_{ij} where R_{ij} is given in the table.

Table of values of R_{ij}
in appropriate units

		$j =$	
		I	II
	1	6	6
$i =$	2	8	5
	3	4	5

All crops need water from irrigation in addition to natural rainfall. The present irrigation system can provide each year 56×10^5 units of water. One acre of crop i grown on land of type j will need W_{ij} units of water each year.

Table of values of W_{ij}

		$j =$	
		I	II
	1	2	3
$i =$	2	3	3
	3	3	2

The total agricultural work force is 7×10^5 men. Every 10 acres of crop 1, crop 2, crop 3 require 2, 1, 3 men, respectively, to be engaged in the various tasks involved in the cultivation during the course of the year. This model leads to the following linear programming problem:

Subject to all the variables being non-negative and satisfying the constraints

$$x_{11} + x_{21} + x_{31} \qquad\qquad\qquad\qquad + u \qquad\qquad\qquad = 14$$
$$x_{12} + x_{22} + x_{32} \quad + v \qquad\qquad\quad = 12$$
$$2x_{11} + 3x_{21} + 3x_{31} + 3x_{12} + 3x_{22} + 2x_{32} \qquad\quad + w \quad = 56$$
$$2x_{11} + x_{21} + 3x_{31} + 2x_{12} + x_{22} + 3x_{32} \qquad\qquad\quad + z = 70$$

Minimise $\qquad R = -6x_{11} - 8x_{21} - 4x_{31} - 6x_{12} - 5x_{22} - 5x_{32}.$

Explain the physical meaning of *all* the variables which appear in this problem. Find a basis, the inverse of this basis, the simplex multipliers associated with this basis and carry out the iterations in the Revised Simplex Method solution of this problem.

The problem was solved using a computer package the final output being:

Basis	Value	Inverse			
x_{21}	4	-2	-2	1	0
x_{11}	10	3	2	-1	0
x_{32}	12	0	1	0	0
z	10	-4	-5	1	1
Simplex Multipliers		2	1	2	0

Objective equals -152.

Interpret and comment on this solution.

A deforestation scheme which would make more land of type I available, together with a major improvement in the irrigation system, is being considered. What is the value of these changes as far as the agricultural economy is concerned?

A fourth crop which is only suitable for cultivation on type II land has been suggested. It would require 2 units of water per acre and is quite labour intensive, requiring 4 men per 10 acres. On current world prices it would yield a net revenue of 6 units per acre. Should it be grown?

9 The Mixxitt Wine Importing Co. imports continental wines for blending and selling in the UK market. It imports three basic red wines which it blends into three table wines sold under different labels. Specifications and other details are as follows.

	Cost/bottle (£)	Available (bottles/year)
French burgundy	1.08	100 000
French claret	0.96	130 000
Spanish red	0.50	150 000

Specifications

Label	Not less than	Not more than	Maximum sales (Bottles/year)	Selling price/ bottle (£)
Beaujolais	30% burgundy	50% Spanish red	200 000	1.96
Nuit St Georges	60% burgundy	30% Spanish red	Unlimited	2.46
St Emelion	60% claret	30% Spanish red	180 000	2.08

Formulate the problem as a linear programming problem and solve it using the Revised Simplex Method. Assume that profit is the company's sole criterion.

10 At the kth iteration in the solution of a linear programming problem using the Revised Simplex Method the *one* basic variable x_p is zero. It had tied with x_r as the variable to leave the basis at the previous iteration. Show that this particular basic feasible solution cannot occur again in a cycle of solutions.

11 Use the Revised Simplex Method to solve some of the transportation problems of Chapter 4. How does it compare with the specialised program for these problems?

12 Use the Revised Simplex Method to solve some of the assignment problems of Chapter 5. How does it compare with the program for Mack's method?

7

Duality in Linear Programming

7.1 The Primal and Dual Problems

Many of our earlier results can be better understood if the concept of duality is introduced. We shall see that to each linear programming problem there corresponds a second (the dual) problem. The relationship between the problems, once understood, enables us to write down the solution to both problems if we have the solution of either problem. First we define our terms.

Corresponding to the **primal** problem:
find $x_j \geqslant 0$ $(j = 1, 2, ..., n)$

such that
$$\sum_{j=1}^{n} a_{ij}x_j \geqslant b_i \quad (i = 1, 2, ..., m) \tag{7.1}$$

which minimise
$$\sum_{j=1}^{n} c_j x_j = z,$$

there corresponds the **dual** problem:
find $y_i \geqslant 0$ $(i = 1, 2, ..., m)$

such that
$$\sum_{i=1}^{m} a_{ij}y_i \leqslant c_j \quad (j = 1, 2, ..., n) \tag{7.2}$$

which maximise
$$\sum_{i=1}^{m} b_i y_i = w.$$

We have in fact already met an example of a primal problem and its dual in Exercises 2, Questions 5 and 6, and again in Exercises 6, Questions 3 and 4.

The primal problem is: find $x_1, x_2 \geqslant 0$ such that

$$x_1 + 3x_2 \geqslant 8$$
$$3x_1 + 4x_2 \geqslant 19$$
$$3x_1 + x_2 \geqslant 7$$

which minimise $\quad 50x_1 + 25x_2 = z.$

The corresponding dual problem is: find $y_1, y_2, y_3 \geqslant 0$ such that

$$y_1 + 3y_2 + 3y_3 \leqslant 50$$
$$3y_1 + 4y_2 + y_3 \leqslant 25$$

which maximise $\quad 8y_1 + 19y_2 + 7y_3 = w.$

The constraints in the primal are of the '$\geqslant$' type, in the dual they are of the '$\leqslant$' type. The number of variables in the dual is the same as the number of constraints in the primal. The matrix of coefficients of the constraints for the dual is the transpose of the matrix of coefficients of the constraints for the primal. The coefficients in the objective function for the dual are the R.H.S. values in the constraints of the primal and vice versa.

In matrix form, the primal problem is:
find $x \geqslant 0$

such that
$$Ax \geqslant b$$

which minimise
$$c^T x = z. \tag{7.3}$$

The dual problem is:
find $y \geqslant 0$

such that
$$A^T y \leqslant c$$

which maximise
$$b^T y = w. \tag{7.4}$$

Although the primal problem has been written to minimise the objective function subject to '$\geqslant$' constraints in non-negative variables (not our previous standard form for an L.P. problem) this represents no loss of generality. Any linear programming problem can be put in this form.

Example 1

Express in standard primal form the problem:
find $x_1, x_2, x_3 \geqslant 0$

such that
$$3x_1 + 4x_2 + x_3 \geqslant 7$$
$$x_1 + 2x_2 + x_3 = 6$$
$$x_3 \leqslant 4$$

which maximise
$$x_1 - 4x_2 - 3x_3 = z'.$$

Hence write down the dual problem.

The maximisation of z' is equivalent to the minimisation of $-z' = z$ say.

The constraint $x_3 \leqslant 4$ is transformed on multiplication by -1 to the constraint $-x_3 \geqslant -4$.

The constraint

$$x_1 + 2x_2 + x_3 = 6$$

is equivalent to the two constraints

$$x_1 + 2x_2 + x_3 \geqslant 6$$
$$x_1 + 2x_2 + x_3 \leqslant 6,$$

i.e.

$$x_1 + 2x_2 + x_3 \geqslant 6$$
$$-x_1 - 2x_2 - x_3 \geqslant -6.$$

Thus our problem may be written:
find $x_1, x_2, x_3 \geqslant 0$

such that
$$3x_1 + 4x_2 + x_3 \geqslant 7$$
$$x_1 + 2x_2 + x_3 \geqslant 6$$
$$-x_1 - 2x_2 - x_3 \geqslant -6$$
$$- x_3 \geqslant -4$$

which minimise
$$-x_1 + 4x_2 + 3x_3 = z,$$

and this is the standard form for the primal.
   Its dual is:
find $y_1, y_2', y_2'', y_3 \geqslant 0$ (you will see the reason for the notation shortly)

such that
$$3y_1 + y_2' - y_2'' + 0y_3 \leqslant -1$$
$$4y_1 + 2y_2' - 2y_2'' + 0y_3 \leqslant 4$$
$$y_1 + y_2' - y_2'' - y_3 \leqslant 3$$

which maximise
$$7y_1 + 6y_2' - 6y_2'' - 4y_3 = w.$$

   This is the standard form for the dual. We see that we can regard $y_2 = y_2' - y_2''$ as a single variable, because of the form of the coefficients (now you see the reason for the notation), and we can write the problem as:
find $y_1, y_2, y_3; y_1, y_3 \geqslant 0$, y_2 unrestricted in sign

such that
$$3y_1 + y_2 \leqslant -1$$
$$4y_1 + 2y_2 \leqslant 4$$
$$y_1 + y_2 - y_3 \leqslant 3$$

which maximise
$$7y_1 + 6y_2 - 4y_3 = w.$$

Example 2
   Find the dual of the problem:
find $x_1 \geqslant 0$ and x_2 unrestricted in sign

such that
$$5x_1 + 3x_2 \geqslant 10$$
$$x_1 - x_2 \leqslant 4$$

which minimise
$$6x_1 + 10x_2.$$

   If we write x_2 as $x_2' - x_2''$ where x_2' and $x_2'' \geqslant 0$ the problem in the standard form for the primal is:
find $x_1, x_2', x_2'' \geqslant 0$

such that
$$5x_1 + 3x_2' - 3x_2'' \geqslant 10$$
$$-x_1 + x_2' - x_2'' \geqslant -4$$

which minimise
$$6x_1 + 10x_2' - 10x_2'' = z.$$
   The corresponding dual is:
find $y_1, y_2 \geqslant 0$

such that
$$5y_1 - y_2 \leqslant 6$$
$$3y_1 + y_2 \leqslant 10$$
$$-3y_1 - y_2 \leqslant -10$$

which maximise
$$10y_1 - 4y_2.$$

This is the standard form for the dual problem. However, the last two constraints are equivalent to

$$3y_1 + y_2 \leqslant 10$$
$$3y_1 + y_2 \geqslant 10,$$

so that the dual could be put in the form:
find $y_1, y_2 \geqslant 0$

such that
$$5y_1 - y_2 \leqslant 6$$
$$3y_1 + y_2 = 10$$

which maximise
$$10y_1 - 4y_2 = w.$$

Thus from the examples of this section we see that to each constraint in the primal there corresponds a variable in the dual, and to each variable in the primal there corresponds a constraint in the dual. If a constraint in the primal is an equation, the corresponding dual variable is not restricted in sign (Example 1); if a primal variable is not restricted in sign the corresponding dual constraint is an equation (Example 2).

7.2   Duality Theorems

I   The dual of the dual problem is the primal problem.

From equations (7.4) the dual problem can be written (in standard form for the primal):
find $y \geqslant 0$

such that $\qquad -A^{\mathrm{T}}y \geqslant -c$
which minimise $\qquad -b^{\mathrm{T}}y = w'.$

Its dual is:
find $x \geqslant 0$

such that $\qquad -Ax \leqslant -b$
which maximise $\qquad -c^{\mathrm{T}}x = z'.$

But this is equivalent to:
find $x \geqslant 0$

such that $\qquad Ax \geqslant b$
which minimise $\qquad c^{\mathrm{T}}x = z$

which is the primal problem.

II   The value of z corresponding to any feasible solution of the primal is greater than or equal to the value of w corresponding to any feasible solution of the dual.

Suppose X and Y are feasible solutions of the primal and dual constraints respectively. Let the corresponding values for the objective functions be

$$Z = c^T X \quad \text{and} \quad W = b^T Y.$$

From equations (7.3) $\quad\quad AX \geqslant b$

Since $\quad\quad\quad\quad\quad\quad Y \geqslant 0$

$$Y^T A X \geqslant Y^T b = b^T Y = W.$$

Also, from equations (7.4)

$$A^T Y \leqslant c,$$

and since $\quad\quad\quad\quad\quad X \geqslant 0$

$$X^T A^T Y \leqslant X^T c = c^T X = Z.$$

But $X^T A^T Y$ is a scalar and so is equal to its transpose $Y^T A X$.

$$\therefore \quad Z = c^T X \geqslant Y^T A X \geqslant b^T Y = W$$

i.e. $\quad\quad\quad\quad\quad\quad\quad\quad\quad Z \geqslant W. \quad\quad\quad\quad\quad\quad\quad\quad (7.5)$

From this result it follows that if there is a feasible value of z that is equal to a feasible value of w, then this must be the minimum value for z and the maximum value of w. Of course it does not follow automatically that such values will exist.

III   If the primal problem has a finite solution with $z = z_{min}$, then the dual problem has a finite solution $w = w_{max} = z_{min}$. Also the simplex multipliers associated with the optimal solution of the primal are the negatives of the variables in the optimal solution of the dual.

If we introduce slack variables into the primal constraints the primal problem takes the form:
find $x_j \geqslant 0$ $(j = 1, 2, ..., n, n + 1, ..., n + m)$

$$
\begin{aligned}
a_{11}x_1 + a_{12}x_2 + \cdots + a_{1n}x_n - x_{n+1} &= b_1 \\
a_{21}x_1 + a_{22}x_2 + \cdots + a_{2n}x_n \quad\quad - x_{n+2} &= b_2 \\
&\cdots\cdots\cdots \\
a_{m1}x_1 + a_{m2}x_2 + \cdots + a_{mn}x_n \quad\quad\quad - x_{n+m} &= b_m
\end{aligned}
$$

which minimise $\quad c_1 x_1 + c_2 x_2 + \cdots + c_n x_n \quad\quad\quad\quad\quad\quad = z.$

If $\pi_1, \pi_2, ..., \pi_m$ are the simplex multipliers for the optimal solution, then on multiplication of the constraints by $\pi_1, ..., \pi_m$ and addition to z we obtain:

$$x_1\left(c_1 + \sum_{i=1}^{m} a_{i1}\pi_i\right) + x_2\left(c_2 + \sum_{i=1}^{m} a_{i2}\pi_i\right) + \cdots + x_n\left(c_n + \sum_{i=1}^{m} a_{in}\pi_i\right)$$

$$- x_{n+1}\pi_1 \quad\quad\quad\quad - x_{n+2}\pi_2 \quad\quad\quad\quad\quad\quad - x_{n+m}\pi_m = z + \sum_{i=1}^{m} \pi_i b_i.$$

$$(7.6)$$

Now, if equation (7.6) represents the optimal canonical form for z, all the coefficients on the L.H.S. will be non-negative.

$$\left.\begin{array}{r} \therefore \quad -\pi_i \geqslant 0 \quad (i = 1, ..., m) \\[2mm] c_j + \sum_{i=1}^{m} a_{ij}\pi_i \geqslant 0 \quad (j = 1, ..., n) \end{array}\right\}. \tag{7.7}$$

and

These conditions are equivalent to

$$\left.\begin{array}{r} -\pi_i \geqslant 0 \quad (i = 1, 2, ..., m) \\[2mm] \sum a_{ij}(-\pi_i) \leqslant c_j \quad (j = 1, ..., n) \end{array}\right\} \tag{7.8}$$

so that the values $-\pi_i$ satisfy the dual constraints.

Of course in equation (7.6) the coefficients of the basic variables will be zero and the non-basic variables themselves are zero so that the L.H.S. is zero and

$$z_{\min} = -\sum_{i=1}^{m} b_i \pi_i = \sum_{i=1}^{m} b_i(-\pi_i). \tag{7.9}$$

Thus $z_{\min} = \sum_{i=1}^{m} b_i(-\pi_i)$. But this is the value of w corresponding to the feasible solution of the dual constraints given by $y_i = -\pi_i$.

Thus, following the remarks after equation (7.5), this must be the maximum value for w. Thus

$$w_{\max} = z_{\min}. \tag{7.10}$$

IV   If the dual problem has a finite solution with $w = w_{\max}$ then the primal problem has a finite solution with $z_{\min} = w_{\max}$. The negatives of the values of the simplex multipliers for the optimal solution of the dual are the values of the variables in the optimal solution of the primal.

This could be deduced from I and III but it may be instructive to give a proof in *matrix notation* along the lines of the proof of III.

The dual with the constraints in equation form can be written

$$A^T y + y_s = c$$

maximise
$$b^T y = w$$

where
$$y_s = \begin{pmatrix} y_{m+1} \\ \vdots \\ y_{m+n} \end{pmatrix} \geqslant 0$$

is the vector of slacks.

If $\rho^T = (\rho_1, \rho_2, ..., \rho_n)$ are the simplex multipliers for the optimal solution of the dual then

$$\rho^T A^T y + \rho^T y_s = \rho^T c = c^T \rho$$
$$\therefore \quad (b^T + \rho^T A^T)y + \rho^T y_s = w + c^T \rho. \tag{7.11}$$

Now since we are *maximising w* (not our usual procedure) the coefficients of y and y_s on the L.H.S. of equation (7.11) must be *non-positive*

$$\left.\begin{array}{r} \therefore \quad \rho^T \leqslant (0, 0, ..., 0) \\[2mm] b^T + \rho^T A^T \leqslant (0, 0, ..., 0) \end{array}\right\} \tag{7.12}$$

i.e.
$$-\rho \geqslant 0$$
$$b + A\rho \leqslant 0$$

which can be written as

$$-\rho \geqslant 0 \qquad (7.13)$$
$$A(-\rho) \geqslant b$$

so that the values $-\rho$ satisfy the primal constraints.

The same arguments as before show that

$$w_{max} = c^T(-\rho) = z_{min}. \qquad (7.14)$$

The consequence of III and IV is that when confronted with a linear programming problem we can choose either to solve the problem as it stands or its dual. If the Revised Simplex Method is used, and this is certainly recommended, both the variables and the simplex multipliers will be obtained. These will allow the simplex multipliers and the variables of the other problem to be determined. This can yield a great saving in computation. It was pointed out in Section 6.4 that the amount of computation in a linear programming problem is related to the number of constraints rather than the number of variables.

Thus in line with the remarks of Section 6.4, if the primal problem has 7 constraints in 3 variables, the transformations will be carried out on a 9×8 matrix in Phase I and an 8×8 matrix in Phase II. The dual problem will have 3 constraints in 7 variables and will not involve a phase I. The transformations will be carried out on a 4×4 matrix. Thus at each iteration the amount of computation is reduced at least four fold.

As an illustration of the results in III and IV consider the problem:
find $x_1, x_2 \geqslant 0$

such that
$$2x_1 + 3x_2 \geqslant 10$$
$$3x_1 + 4x_2 \geqslant 19$$
$$x_1 + 2x_2 \geqslant 9$$

which minimise
$$7x_1 + 10x_2 = z.$$

Its dual is:
find $y_1, y_2, y_3 \geqslant 0$

such that
$$2y_1 + 3y_2 + y_3 \leqslant 7$$
$$3y_1 + 4y_2 + 2y_3 \leqslant 10$$

which maximise
$$10y_1 + 19y_2 + 9y_3 = w.$$

The computer solution of the first problem shows that in the optimal solution $x_1 = 1$, $x_2 = 4$, $\pi_1 = 0$, $\pi_2 = -2$, $\pi_3 = -1$ (from the last row) with $z_{min} = 47$. Thus for the second problem the solution is $y_1 = 0$, $y_2 = 2$, $y_3 = 1$ with $\rho_1 = -1$, $\rho_2 = -4$ with $w_{max} = 47$.

This is confirmed by the computer solution of the second problem. A little care is needed here. In line with our treatment of linear programming problems we have minimised

$$w' = -10y_1 - 19y_2 - 9y_3$$

in the computer solution, and this has reversed the signs of the objective function and the simplex multipliers.

REVISED SIMPLEX METHOD

PHASE I

INITIAL TABLEAU

BASE VAR.	VALUE	X1	X2	X3	X4	X5	X6	X7	X8
X6	10.00	2.00	3.00	−1.00	0.00	0.00	1.00	0.00	0.00
X7	19.00	3.00	4.00	0.00	−1.00	0.00	0.00	1.00	0.00
X8	9.00	1.00	2.00	0.00	0.00	−1.00	0.00	0.00	1.00
OBJECTIVE Z		7.00	10.00	0.00	0.00	0.00	0.00	0.00	0.00
ARTIFICIAL W		0.00	0.00	0.00	0.00	0.00	1.00	1.00	1.00

ITERATION 0

D1	D2	D3	D4	D5	D6	D7	D8
−6.00	−9.00	1.00	1.00	1.00	0.00	0.00	0.00

X2 ENTERS BASIS

BASE VAR.	VALUE	INVERSE OF BASIS			A′[I]
X6	10.00	1.00	0.00	0.00	3.00
X7	19.00	0.00	1.00	0.00	4.00
X8	9.00	0.00	0.00	1.00	2.00
OBJ SIM MULT	0.00	0.00	0.00	0.00	10.00
ART SIM MULT	−38.00	−1.00	−1.00	−1.00	−9.00

X6 FROM CONSTRAINT 1 LEAVES THE BASIS

ITERATION 1

D1	D2	D3	D4	D5	D6	D7	D8
0.00	0.00	−2.00	1.00	1.00	3.00	0.00	0.00

X3 ENTERS BASIS

BASE VAR.	VALUE	INVERSE OF BASIS			A′[I]
X2	3.33	0.33	0.00	0.00	−0.33
X7	5.67	−1.33	1.00	0.00	1.33
X8	2.33	−0.67	0.00	1.00	0.67
OBJ SIM MULT	−33.33	−3.33	0.00	0.00	3.33
ART SIM MULT	−8.00	2.00	−1.00	−1.00	−2.00

X8 FROM CONSTRAINT 3 LEAVES THE BASIS

ITERATION 2

D1	D2	D3	D4	D5	D6	D7	D8
−1.00	0.00	0.00	1.00	−2.00	1.00	0.00	3.00

X5 ENTERS BASIS

BASE VAR.	VALUE	INVERSE OF BASIS			A′[I]
X2	4.50	0.00	0.00	0.50	−0.50
X7	1.00	0.00	1.00	−2.00	2.00
X3	3.50	−1.00	0.00	1.50	−1.50
OBJ SIM MULT	−45.00	0.00	0.00	−5.00	5.00
ART SIM MULT	−1.00	0.00	−1.00	2.00	−2.00

X7 FROM CONSTRAINT 2 LEAVES THE BASIS

PHASE I SUCCESSFUL

PHASEII

ITERATION 3
```
     C1      C2      C3      C4      C5
   -0.50    0.00    0.00    2.50    0.00
```
X1 ENTERS BASIS

BASE VAR.	VALUE	INVERSE OF BASIS			A^[I]
X2	4.75	0.00	0.25	0.00	0.75
X5	0.50	0.00	0.50	-1.00	0.50
X3	4.25	-1.00	0.75	0.00	0.25
OBJ SIM MULT	-47.50	0.00	-2.50	0.00	-0.50

X5 FROM CONSTRAINT 2 LEAVES THE BASIS

ITERATION 4
```
     C1      C2      C3      C4      C5
    0.00    0.00    0.00    2.00    1.00
```

BASE VAR.	VALUE	INVERSE OF BASIS		
X2	4.00	0.00	-0.50	1.50
X1	1.00	0.00	1.00	-2.00
X3	4.00	-1.00	0.50	0.50
OBJ SIM MULT	-47.00	0.00	-2.00	-1.00

FINAL SOLUTION

MINIMUM OF Z =   47.00

CONSTRAINT	BASIS	VALUE	STATE	SLACK	PI
1	2	4.00	SLACK	4.00	0.00
2	1	1.00	BINDING	0.00	-2.00
3	3	4.00	BINDING	0.00	-1.00

REVISED SIMPLEX METHOD

THERE IS NO PHASE I

INITIAL TABLEAU

BASE VAR.	VALUE	X1	X2	X3	X4	X5
X4	7.00	2.00	3.00	1.00	1.00	0.00
X5	10.00	3.00	4.00	2.00	0.00	1.00
OBJECTIVE	Z	-10.00	-19.00	-9.00	0.00	0.00

ITERATION 0
```
     C1      C2      C3      C4      C5
  -10.00  -19.00   -9.00    0.00    0.00
```
X2 ENTERS BASIS

BASE VAR.	VALUE	INVERSE OF BASIS		A^[I]
X4	7.00	1.00	0.00	3.00
X5	10.00	0.00	1.00	4.00
OBJ SIM MULT	0.00	0.00	0.00	-19.00

X4 FROM CONSTRAINT 1 LEAVES THE BASIS

```
ITERATION   1
    C1      C2      C3      C4      C5
   2.67    0.00   -2.67    6.33    0.00
X3 ENTERS BASIS
   BASE VAR.       VALUE INVERSE OF BASIS   A´[I]
      X2            2.33    0.33    0.00    0.33
      X5            0.67   -1.33    1.00    0.67
OBJ SIM MULT       44.33    6.33    0.00   -2.67
X5 FROM CONSTRAINT 2 LEAVES THE BASIS

ITERATION   2
    C1      C2      C3      C4      C5
   4.00    0.00    0.00    1.00    4.00
   BASE VAR.       VALUE INVERSE OF BASIS
      X2            2.00    1.00   -0.50
      X3            1.00   -2.00    1.50
OBJ SIM MULT       47.00    1.00    4.00

FINAL SOLUTION

MINIMUM OF Z =   -47.00

CONSTRAINT    BASIS      VALUE     STATE       SLACK        PI
    1           2        2.00     BINDING      0.00       1.00
    2           3        1.00     BINDING      0.00       4.00
```

One further point is worth mentioning. From equation (7.6), in the optimal solution of the primal

$$\pi_i x_{n+i} = 0 \quad (i = 1, \ldots, m). \tag{7.15}$$

From equation (7.11), in the optimal solution of the dual

$$\rho_j y_{m+j} = 0 \quad (j = 1, \ldots, n). \tag{7.16}$$

Thus at least one of π_i and x_{n+i} is zero and at least one of ρ_j and y_{m+j} is zero. These results may be stated in the following form.

If in the optimal solution of the primal the kth constraint is not binding ($x_{n+k} > 0$), then the kth dual variable ($-\pi_k$) is zero. If the kth dual variable is positive then the kth primal constraint is binding. These results are sometimes known as the complementary slackness principle.

7.3   Looking Back in the Light of Duality

The ideas of duality can help us to understand some of the earlier results for linear programming.

The algorithm for the Dual Simplex Method (Section 3.3) was derived without recourse to duality. However, this does throw light on the procedure.

Consider the problem:
find $x_1, x_2, x_3 \geqslant 0$

such that

$$x_1 \quad\;\;\; + \; 3x_3 \geqslant 3$$
$$x_2 + \; 2x_3 \geqslant 5$$

which minimise

$$4x_1 + 6x_2 + \; 18x_3 = z.$$

Because all the coefficients in z are positive we can avoid the introduction of artificial variables and solve the problem using the Dual Simplex Method. The successive tableaux follow.

Iteration	Basis	Value	x_1	x_2	x_3	x_4	x_5
0	x_4	-3	-1	0	-3	1	$\cdot$
	x_5	-5	0	-1^*	-2	$\cdot$	1
	$-z$	0	4	6	18	$\cdot$	$\cdot$
1	x_4	-3	-1	$\cdot$	-3^*	1	0
	x_2	5	0	1	2	$\cdot$	-1
	$-z$	-30	4	$\cdot$	6	$\cdot$	6
2	x_3	1	$\frac{1}{3}$	$\cdot$	1	$-\frac{1}{3}$	0
	x_2	3	$-\frac{2}{3}$	1	$\cdot$	$\frac{2}{3}$	-1
	$-z$	-36	2	$\cdot$	$\cdot$	2	6

Thus in the optimal solution $x_1 = 0$, $x_2 = 3$, $x_3 = 1$ with $z_{min} = 36$.
The simplex multipliers (coefficients of the slacks x_4 and x_5 in the final form for z) are $\pi_1 = 2$ and $\pi_2 = 6$.
Consider the dual problem:
find $y_1, y_2 \geqslant 0$

such that

$$y_1 \quad\quad\;\; \leqslant \; 4$$
$$y_2 \leqslant \; 6$$
$$3y_1 + 2y_2 \leqslant 18$$

which maximise

$$3y_1 + \; 5y_2 = w.$$

To conform with our standard approach we minimise

$$w' = -3y_1 - 5y_2.$$

The successive tableaux of the Simplex Method solution follow.

Iteration	Basis	Value	y_1	y_2	y_3	y_4	y_5
	y_3	4	1	0	1	.	.
0	y_4	6	0	1^*	.	1	.
	y_5	18	3	2	.	.	1
	$-w'$	0	-3	-5	.	.	.
	y_3	4	1	.	1	0	.
1	y_2	6	0	1	.	1	.
	y_5	6	3^*	.	.	-2	1
	$-w'$	30	-3	.	.	5	.
	y_3	2	.	.	1	$\frac{2}{3}$	$-\frac{1}{3}$
2	y_2	6	.	1	.	1	0
	y_1	2	1	.	.	$-\frac{2}{3}$	$\frac{1}{3}$
	$-w'$	36	.	.	.	3	1

The simplex multipliers (for w' as objective function) are $\rho_1 = 0$, $\rho_2 = 3$, $\rho_3 = 1$ (the coefficients of the slacks y_3, y_4 and y_5). These give the values of the primal variables. The dual variables $y_1 = 2$ and $y_2 = 6$ are the simplex multipliers for the primal. The calculations that have been carried out, the Dual Simplex Method for the primal and the Simplex Method for the dual are virtually identical.

The 'Stepping Stones' algorithm for the transportation problem is in fact an example where we choose to solve the dual rather than the primal problem.

The transportation problem can be put in the form:
find $x_{ij} \geqslant 0$

such that $\quad x_{11} + x_{12} + \cdots + x_{1n} \qquad\qquad\qquad\qquad = a_1$

$\qquad\qquad\qquad x_{21} + \qquad\qquad + x_{2n} \qquad\qquad = a_2$

. . . .

$\qquad\qquad\qquad\qquad\qquad\qquad\qquad x_{m1} + \cdots + x_{mn} = a_m$

$\quad x_{11} \qquad\qquad\qquad + x_{21} \qquad\qquad + x_{m1} \qquad\qquad = b_1$

$\qquad\quad x_{12} \qquad\qquad\qquad + x_{22} \qquad\qquad + x_{m2} \qquad = b_2$

. . . .

$\qquad\qquad\quad x_{1n} \qquad\qquad + x_{2n} \qquad\qquad + x_{mn} = b_m$

and $\qquad c_{11}x_{11} + c_{12}x_{12} + \qquad\qquad\qquad + c_{mn}x_{mn} = z$

is to be minimised (see equations (4.2)).

For the dual problem we have to find u_i and v_j which are not restricted in sign, since the primal constraints are equations, such that

$$
\begin{array}{llll}
u_1 & & + v_1 & \leqslant c_{11} \\
& u_2 & + v_2 & \leqslant c_{12} \\
& u_i & + v_j & \leqslant c_{ij} \\
& u_m & + v_n & \leqslant c_{mn}
\end{array}
$$

and $\qquad a_1 u_1 + a_2 u_2 + \quad a_m u_m + b_1 v_1 + \qquad b_n v_n = w$

is to be maximised.

We try to satisfy all the dual constraints $u_i + v_j \leqslant c_{ij}$. This is not too difficult since each constraint only involves two variables. In fact in the $m + n - 1$ cells corresponding to the basic variables in the primal problem we achieve equality in the dual constraints (complementary slackness). When the dual constraints are satisfied as inequalities for the other cells we shall have a solution to both problems.

For, in the primal problem, multiplication of the ith row and jth column by $-u_i$ and $-v_j$ followed by addition to z gives:

$$
z - \sum_{i=1}^{m} a_i u_i - \sum_{j=1}^{n} b_j v_j = \sum_i \sum_j (c_{ij} - u_i - v_j) x_{ij} = z - w
$$

(see equation (4.7)).

We choose u_i and v_j so that $u_i + v_j = c_{ij}$ for the non-zero x_{ij}. Thus we can ensure $z = w$ at each stage.

When for the non-basic x_{ij}, $c_{ij} - u_i - v_j \geqslant 0$ we have a solution of the dual constraints with equal values for z and w. Hence this must give the optimal solution for both problems.

The reader might ask what happens if we treat the primal problem, or indeed any linear programming problem, according to the general (Kuhn–Tucker) theory for constrained optimisation. In case the reader is not familiar with these ideas we develop them from first principles in the treatment that follows.

Thus we consider the problem of minimising

$$
f(x) = \sum_{j=1}^{n} c_j x_j \tag{7.17}
$$

Subject to $\qquad x_j \geqslant d_j \quad (j = 1, ..., n) \tag{7.18}$

and $\qquad \sum_{j=1}^{n} a_{ij} x_j \geqslant b_i \quad (i = 1, ..., m). \tag{7.19}$

Normally the d_j are all zero but it is convenient to generalise the problem at this stage.

We write the constraints (7.18) and (7.19) in the form

$$
x_j - u_j^2 - d_j = 0 \tag{7.20}
$$

and $\qquad \sum_{j=1}^{n} a_{ij} x_j - v_i^2 - b_i = 0 \tag{7.21}$

and define the Lagrange function

$$F(x,\pi,\sigma,u,v) = \sum_{j=1}^{n} c_j x_j + \sum_{i=1}^{m} \pi_i \left(\sum_{j=1}^{n} a_{ij} x_j - v_i^2 - b_i \right) + \sum_{j=1}^{n} \sigma_j (x_j - u_j^2 - d_j) \quad (7.22)$$

where the π_i and σ_j are the so-called Lagrange multipliers.

The constrained minimum of equation (7.17) will be given by the unconstrained minimum of equation (7.22). The necessary conditions for the latter are given by

$$\frac{\partial F}{\partial x_j} = 0, \quad \text{i.e.} \quad c_j + \sum_{i=1}^{m} a_{ij} \pi_i + \sigma_j = 0; \quad j = 1, \ldots, n \quad (7.23)$$

$$\frac{\partial F}{\partial \pi_i} = 0, \quad \text{i.e.} \quad \sum_{j=1}^{n} a_{ij} x_j - v_i^2 - b_i = 0; \quad i = 1, \ldots, m \quad (7.24)$$

$$\frac{\partial F}{\partial \sigma_j} = 0, \quad \text{i.e.} \quad x_j - u_j^2 - d_j = 0; \quad j = 1, \ldots, n \quad (7.25)$$

$$\frac{\partial F}{\partial v_i} = 0, \quad \text{i.e.} \quad \pi_i v_i = 0$$

which implies $\pi_i \left(\sum_{j=1}^{n} a_{ij} x_j - b_i \right) = 0; \quad i = 1, \ldots, m \quad (7.26)$

$$\frac{\partial F}{\partial u_j} = 0, \quad \text{i.e.} \quad \sigma_j u_j = 0$$

which implies $\sigma_j (x_j - d_j) = 0; \quad j = 1, \ldots, n. \quad (7.27)$

We recognise that equations (7.24) and (7.25) are equivalent to equations (7.18) and (7.19). Equations (7.26) and (7.27) reflect the complementary slackness principle. The Lagrange multipliers π_i are equivalent to the simplex multipliers.

Of course for values of the variables satisfying the constraints

$$F(\ldots) = f(\cdot)$$

and

$$F_{min} = f_{min}. \quad (7.28)$$

Now from equation (7.22)

$$\frac{\partial F_{min}}{\partial b_i} = -\pi_i \quad \text{and} \quad \frac{\partial F_{min}}{\partial d_j} = -\sigma_j. \quad (7.29)$$

Compare this with equation (3.17).

Now as b_i or d_j are increased the constrained region is made smaller so that F_{min} cannot decrease.

$$\therefore \quad -\pi_i \geqslant 0 \quad \text{and} \quad -\sigma_j \geqslant 0. \quad (7.30)$$

Compare this with equations (7.7). Thus we have from equation (7.23)

$$c_j + \sum_{i=1}^{m} a_{ij}\pi_i = -\sigma_j \geqslant 0.$$

$$\therefore \quad -\pi_i \geqslant 0 \quad \text{and} \quad \sum_{i=1}^{m} a_{ij}(-\pi_i) \leqslant c_j$$

so that the values for $-\pi_i$ satisfy the dual constraints.

Now for the value of F_{min} corresponding to the values for x, π etc. which satisfy the necessary conditions

$$F_{min} = \Sigma\, c_j x_j$$

$$= \sum_{j=1}^{n} x_j \left(c_j + \sum_{i=1}^{m} a_{ij}\pi_i \right) - \Sigma\, \pi_i b_i$$

$$= -\sum_{i=1}^{n} x_j \sigma_j + \Sigma\, b_i(-\pi_i)$$

$$= \Sigma\, b_i(-\pi_i) \quad \text{because of equation (7.27) when } d_j = 0.$$

Thus, for these values

$$\Sigma\, c_j x_j = \Sigma\, b_i(-\pi_i), \tag{7.31}$$

i.e. $z_{min} = w_{max}$ in our primal dual notation.

Thus the Kuhn–Tucker theory for constrained optimisation tells us that the necessary conditions for the solution of the primal are equivalent to finding the solution of the dual. It does not in itself provide us with a practical solution method for either problem. For that we need a procedure such as the Simplex Method or the Revised Simplex Method.

Exercises 7

1   Write down the dual of the problem:
find $x_1, x_2 \geqslant 0$

such that
$$\begin{aligned} 3x_1 + \ 5x_2 &\geqslant 18 \\ x_1 + \ 9x_2 &\geqslant 30 \\ 2x_1 + \ 7x_2 &\geqslant 27 \end{aligned}$$
which minimise $11x_1 + 44x_2.$

2   Write down the dual of the problem:
find $x_1, x_2 \geqslant 0$ and x_3 which is not restricted in sign

such that
$$\begin{aligned} 2x_1 + \ x_2 + \ 4x_3 &\geqslant 22 \\ x_1 + \ x_2 + \ x_3 &= \ 9 \\ x_1 + 2x_2 + \ 3x_3 &\leqslant 18 \end{aligned}$$
which minimise $5x_1 + 7x_2 + 13x_3.$

3 Show that if the kth variable in the primal problem is not restricted in sign, then the kth constraint in the dual problem is an equation constraint.

4 Show that if the kth constraint in the primal problem is an equation then the kth dual variable is unrestricted in sign.

5 Find $x_1, x_2, x_3 \geqslant 0$

such that
$$2x_1 + 3x_2 + x_3 \geqslant 11$$
$$x_1 + 2x_2 + 5x_3 \geqslant 20$$
$$3x_1 + x_2 + 2x_3 \geqslant 11$$
which minimise $\quad 11x_1 + 14x_2 + 15x_3 = z.$

Write down and solve the dual problem and verify the correctness of the duality theorems III and IV.

6 For the problem:
find $x_1, x_2 \geqslant 0$

such that
$$x_1 - x_2 \geqslant 1$$
$$- x_2 \geqslant -2$$
which minimise $\quad - x_1 - x_2 = z.$

(Example 3 of Section 1.2), write down the dual and show that the constraints have no feasible solution.

7 Show that if the primal (dual) problem has an unbounded solution with an unbounded value for z (w), then the dual (primal) problem has no feasible solution.

8 Show that the converse of the results stated in Question 7 is not true. Construct a counter example.

9 Verify theorems III and IV for the problem:
find $x_1, x_2, x_3 \geqslant 0$

such that
$$x_1 + 5x_2 + x_3 \geqslant 7$$
$$2x_1 + x_2 + 3x_3 \geqslant 9$$
$$3x_1 + 2x_2 + 5x_3 \geqslant 14$$
which minimise $\quad 7x_1 + 4x_2 + 11x_3 \quad = z.$

10 A bakery sells cakes which spoil quickly and cannot be sold except on the day of baking. The demand probabilities are known to be $p_0 = \frac{1}{8}$, $p_1 = \frac{1}{8}$, $p_2 = \frac{1}{4}$, $p_3 = \frac{1}{2}$, where p_j is the probability that the demand on a day is j. The cost of baking j cakes in a day is Cj where C is a constant. Show that the problem of satisfying the demand on average at least one day in four while minimising baking costs can be expressed as a linear programming problem.

By finding a dual of this problem which can be solved graphically, show that the minimum expected cost is given by baking three cakes on one day in seven and doing no further baking.

8
Linear Programming in Integers

8.1 Some Problems involving Integer Programming

We have seen in section 4.1 that in some problems, physical considerations restrict our variables to integer values. In that problem the number of beds moved from a warehouse to a store had to be a whole number. Care has to be exercised in the matter however, for in Example 1 of section 1.1 it may well have been possible to produce $23\frac{1}{4}$ kits for example in one week, the implication being that the work left unfinished on a kit at the end of one week could be completed at the beginning of the next. Of course, in such problems, each one has to be examined carefully in the context in which it arises.

However, we have also seen that in transportation and assignment problems, any restriction to integer valued variables will automatically be satisfied by the solution derived by the algorithm. We do not have to take any special steps to ensure integer solutions in such problems. However, this will not be the case in all linear programming problems in which the variables are additionally restricted to integer values. In general such a limitation which is not a linear constraint requires special consideration.

Before we consider how this can be done, we give some examples of some problems which can be formulated as linear programming problems, in which the variables are also restricted to integer values.

Example 1   The shortest route through a network.

Suppose a network consists of N nodes, $1, 2, 3, ..., N$, (see Fig. 8.1) and suppose we wish to find the shortest route from node 1 to node N, by way of the linked nodes.

Suppose the distance from node i to node j is d_{ij}. If these two nodes are not linked then d_{ij} can be taken to be infinite. To find the shortest route from 1 to N we proceed as follows. Let $x_{ij} = 1$ if the branch from i to j is on the route; let $x_{ij} = 0$ if the branch from i to j is not on the route.

Then for each i, on the route or not, $\sum_j x_{ij} \leqslant 1$ ($= 1$ for $i = 1$)

and for each j, on the route or not, $\sum_i x_{ij} \leqslant 1$ ($= 1$ for $j = N$) $^{(8.1)}$

If we arrive at a point we must leave it, so for each j, except 1 and N,

$$\sum_i x_{ij} = \sum_k x_{jk}. \tag{8.2}$$

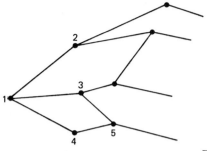

Figure 8.1

Subject to (8.1) and (8.2) we minimise

$$D = \sum_i \sum_j x_{ij} d_{ij}. \tag{8.3}$$

The 0, 1 values for the x_{ij} can be ensured by the constraint

$$0 \leqslant x_{ij} \leqslant 1 \quad \text{and} \quad x_{ij} \text{ is an integer.} \tag{8.4}$$

Thus our problem is to minimise (8.3) which is a linear function of the x_{ij} subject to (8.1), (8.2) and (8.4), all the constraints being linear, but with the additional requirement that all the x_{ij} are integers.

Example 2

Fig. 8.2 shows the floor plan of an art gallery, the various rooms being connected by the open doors as shown. Find the smallest number of attendants, who if placed in the doorways can supervise all the rooms.

There are 6 doorways A, B, ..., F and 5 rooms 1, ..., V.

$$\text{Let } x_a = 1 \text{ if there is an attendant at A}$$
$$= 0 \text{ otherwise etc.}$$

Then since each room has be supervised:

$$
\begin{array}{llll}
x_a + x_b & & \geqslant 1 & \text{(Room I)} \\
x_a + x_c & & \geqslant 1 & \text{(Room II)} \\
x_b + x_c + x_d + x_e & \geqslant 1 & \text{(Room III)} \\
x_d + x_f & & \geqslant 1 & \text{(Room IV)} \\
x_e + x_f & & \geqslant 1 & \text{(Room V)}
\end{array} \tag{8.5}
$$

Subject to (8.5) and $\qquad 0 \leqslant x_i \leqslant 1 \ (i = a, b, ..., f) \tag{8.6}$

and x_i is an integer we have to minimise

$$N = x_a + x_b + x_c + x_d + x_e + x_f. \tag{8.7}$$

Thus we have a linear programming problem where all the variables have to be integers.

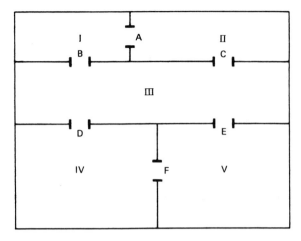

Figure 8.2

8.2 Gomory's Method for all Integer Linear Programming

Suppose we have a linear programming problem in which *all* the variables are confined to integer values. We first solve the problem by the Simplex Method (say) ignoring the integer requirement. If in our optimum solution all the variables happen to assume integer values then we have the solution and there is nothing more to do. Otherwise we have to adopt further measures.

The idea underlying Gomory's method is to express the all integer requirement as a further linear constraint, or more often as a sequence of further linear constraints. The enlarged problem(s) including these extra constraints has to be solved until an all integer solution is obtained. Gomory's method gives a systematic procedure for introducing the extra constraints.

Suppose that we have solved our problem by the Simplex Method and that the optimal simplex tableau is as shown.

Basis	Value	x_1	x_2		x_m	x_{m+1}			x_n
x_1	b_1	1	.		.	a_{1m+1}			a_{1n}
x_2	b_2	.	1		.	a_{2m+1}			a_{2n}
			.	.	.				
x_i	b_i	.	.		.	a_{im+1}			a_{in}
x_m	b_m	.	.		1	a_{mm+1}			a_{mn}
$-z$	$-z_0$	.	.		.	c_{m+1}		c_{n-1}	c_n

Suppose that b_i is not an integer so that our solution does not satisfy the all integer requirement. In this row we split every coefficient into its largest lower integer and a *non-negative* fraction. (i.e. $3\frac{1}{8} = 3 + \frac{1}{8}$, $-1\frac{1}{4} = -2 + \frac{3}{4}$ etc.)

$$\text{Thus } b_i = n_{i0} + f_{i0}, \quad a_{ij} = n_{ij} + f_{ij}$$

where $$0 \leqslant f_{ij} < 1 \quad \text{for } j = 0, \, m+1, \, m+2, \ldots, n.$$

Let $$k_i = f_{im+1}x_{m+1} + f_{im+2}x_{m+2} + \cdots + f_{in}x_n \geqslant 0. \tag{8.8}$$

For row i of the optimal tableau we have

$$n_{i0} + f_{i0} = x_i + (n_{im+1} + f_{im+1})x_{m+1} + \cdots + (n_{in} + f_{in})x_n \tag{8.9}$$

so that by subtraction we obtain

$$f_{i0} - k_i = x_i + n_{im+1}x_{m+1} + n_{im+2}x_{m+2} + \cdots + n_{in}x_n - n_{i0}. \tag{8.10}$$

Now since all our variables must be integers the R.H.S. of (8.10) must be an integer and it cannot be larger than f_{i0} since $k_i \geqslant 0$. But this is a fraction, so the R.H.S. cannot exceed zero.

Thus $$f_{i0} - k_i \leqslant 0, \quad \text{i.e.} \quad k_i \geqslant f_{i0}.$$

Thus the integer requirement is equivalent to

$$k_i \geqslant f_{i0}$$

i.e.

$$-f_{im+1}x_{m+1} - f_{im+2}x_{m+2} - \cdots - f_{in}x_n + x_{n+1} = -f_{i0} \tag{8.11}$$

where x_{n+1} is a new variable which must be *non-negative and an integer* since it equals $k_i - f_{i0}$ (see (8.10)).

Equation (8.11) is the additional constraint we need to add to the problem. At this stage the value of x_{n+1} is negative. However, since the objective function in its current form will have all positive coefficients we can use the Dual Simplex Method to proceed with the solution of the new enlarged problem.

Since whenever we want to remove a non-integer value we introduce a new variable and constraint it is not immediately obvious that the process will terminate in a finite number of iterations. The proof that the method will terminate is not trivial and is not included here. The interested reader should consult Gomory's paper.

Example 1
Find non-negative integers x_1, x_2 such that

$$x_1 + x_2 \leqslant 5$$
$$x_1 + 5x_2 \leqslant 10$$

which minimise $$-2x_1 - 9x_2 = z.$$

We introduce slack variables and solve the problem in the usual way, ignoring the integer condition. The successive simplex tableaux are shown.

Basis	Value	x_1	x_2	x_3	x_4
x_3	5	1	1	1	.
x_4	10	1	5^*	.	1
$-z$	0	-2	-9	.	.
x_3	3	$\frac{4}{5}^*$	.	1	$-\frac{1}{5}$
x_2	2	$\frac{1}{5}$	1	.	$\frac{1}{5}$
$-z$	18	$-\frac{1}{5}$	.	.	$\frac{9}{5}$
x_1	$\frac{15}{4}$	1	.	$\frac{5}{4}$	$-\frac{1}{4}$
x_2	$\frac{5}{4}$	.	1	$-\frac{1}{4}$	$\frac{1}{4}$
$-z$	$\frac{75}{4}$	.	.	$\frac{1}{4}$	$\frac{7}{4}$

It is clear that in the optimal solution the values of x_1 and x_2 are not integers. Equation (8.9) for the first row takes the form

$$3 + \tfrac{3}{4} = x_1 + (1 + \tfrac{1}{4})x_3 + (-1 + \tfrac{3}{4})x_4$$

whence (8.11) becomes

$$-\tfrac{1}{4}x_3 - \tfrac{3}{4}x_4 + x_5 = -\tfrac{3}{4} \qquad (8.12)$$

where x_5 is also a non-negative integer variable. This is the additional constraint we add to the problem. Because the objective function is in optimal form we can proceed using the Dual Simplex Method.

Basis	Value	x_1	x_2	x_3	x_4	x_5
x_1	$\frac{15}{4}$	1	.	$\frac{5}{4}$	$-\frac{1}{4}$	.
x_2	$\frac{5}{4}$	.	1	$-\frac{1}{4}$	$\frac{1}{4}$	.
x_5	$-\frac{3}{4}$	.	.	$-\frac{1}{4}^*$	$-\frac{3}{4}$	1
$-z$	$\frac{75}{4}$	.	.	$\frac{1}{4}$	$\frac{7}{4}$	.
x_1	0	1	.	.	-4	5
x_2	2	.	1	.	1	-1
x_3	3	.	.	1	3	-4
$-z$	18	.	.	.	1	1

The integer solution appears after just one iteration. Of course had this not occurred it would have been necessary to add an additional constraint, corresponding to a non-integer basic variable, to the problem etc. However that is not so for our problem. The optimal solution is given by $x_1 = 0$, $x_2 = 2$ and $z = -18$.

It is important to note that the integer optimal solution cannot be derived from the optimal non-integer solution simply by rounding the variables to the nearest integer value. Of course this problem is capable of a two dimensional graphical

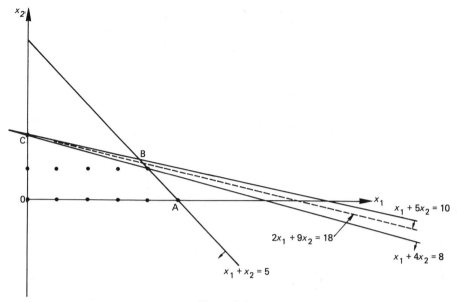

Figure 8.3

representation as shown in Fig. 8.3. The integer points of the constrained region are shown. The optimal point for the non-integer problem lies at B; for the integer problem at C.

In terms of the variables x_1, x_2 the additional constraint (8.12) takes the form

$$x_5 = -\tfrac{3}{4} + \tfrac{1}{4}x_3 + \tfrac{3}{4}x_4 \geqslant 0$$

i.e.
$$-\tfrac{3}{4} + \tfrac{1}{4}(5 - x_1 - x_2) + \tfrac{3}{4}(10 - x_1 - 5x_2) \geqslant 0$$

i.e.
$$x_1 + 4x_2 \leqslant 8.$$

This additional constraint is also shown in Fig. 8.3. It cuts a slice off the con-strained region in such a way that the optimal solution for the enlarged problem lies at an integer point. This indeed is the geometric interpretation to be placed on Gomory's method. In order to satisfy the integer condition we add further constraints ('cutting planes') to the problem. These remove parts of the original constrained region in such a way that ultimately the optimal solution occurs at an integer point.

8.3   A Computer Program for Gomory's Method

The program that follows implements Gomory's method. The intermediate out-put, which can be suppressed if only the final solution is required, follows the pattern of a hand calculation.

As written, the program allows up to 20 extra constraints to be added. This is included in the constant *mmax* and could of course be increased. The program first solves the problem by the Simplex Method, ignoring the integer requirement. If necessary it introduces an artificial objective function in order to carry out this procedure. Then additional constraints are introduced according to Gomory's procedure, and the Dual Simplex Method used to carry out the subsequent minimisation.

The program Gomory is based on the program *FullSimplex* of Chapter 2. The functions *posfrac* and *integersolution*, as well as the procedure *addconstraint*, are added. The procedure *Simplexmod* is a modified version of the procedure *Simplex* in *FullSimplex*, containing the addition of the Dual Simplex Method in the procedure *Nextbasicvariable*. The Main Program is similar to that of *FullSimplex*, additions being the initialisation of *m* to *mstart* and iterative calls to *Simplexmod* in Phase II until an integer solution is found. The program listing given below shows only those parts of *Gomory* that differ from *FullSimplex*. The values in the CONST part are appropriate for Example 1 of Section 8.2 and the data file would contain the following values:

$$1\ 0\ 0\ 2\quad 1\ 1\ 5\quad 1\ 5\ 10\quad -2\ -9$$

The full output for that problem follows the program listing.

```
PROGRAM Gomory (input,output);
CONST
   nvar=2;                { No. of variables }                              {**}
   mstart=2;              { Initial no. of constraints }                    {**}
   mmax=22;               { Max. no. of constraints, set at mstart+20 }     {**}
   ncolsmax=26;           { Max. n. of tableau columns, set at 3*nvar+20 }  {**}
   fwt=7;  dpt=2;         { Output format constants for tableau values }    {**}
   fwi=1;                 { Output format constant for indices }            {**}
   largevalue = 1.0E20;   smallvalue=1.0E-10;                               {**}

TYPE   { Definitions as for Fullsimplex }

VAR    { Declarations as for FullSimplex with addition of m }

   m : integer;              { No. of constraints in current tableau }


PROCEDURE inputdata;  { See FullSimplex }

PROCEDURE initialise;  { See FullSimplex }

PROCEDURE completetableau;  { See FullSimplex }

PROCEDURE outputtableau (p:phase);  { See FullSimplex }

FUNCTION posfrac (x:real): real;
{ Find fraction f, where x=n+f, n an integer and 0<=f<1 }
VAR y : real;
BEGIN
  IF abs(x-round(x))<smallvalue THEN posfrac:=0.0
  ELSE
  BEGIN  y := abs(x-trunc(x));
    IF x>=0.0 THEN posfrac:=y ELSE posfrac:=1.0-y
  END
END; { posfrac }
```

```
PROCEDURE addconstraint (k:integer);
{ Add an extra constraint to the tableau, based on (8.11) }
VAR  i, j : integer;
BEGIN  m:=m+1;  n2:=n2+1;  nonbasic[n2]:=false;  basic[m]:=n2;
  b[m]:= -posfrac(b[k]);  a[m,n2]:=1.0;  c[n2]:=0.0;
  FOR j:=1 TO n2-1 DO
    IF nonbasic[j] THEN a[m,j]:= -posfrac(a[k,j])
    ELSE a[m,j]:=0.0;
  FOR i:=1 TO m-1 DO a[i,n2]:=0.0;  it := it+1
END; { addconstraint }


FUNCTION integersolution: boolean;
{ Test whether the current ´solution´ is the required integer solution }
VAR  i : integer;  stillinteger : boolean;
BEGIN  i:=0;  stillinteger:=true;
  WHILE (i<m) AND stillinteger DO
  BEGIN  i:=i+1;
    IF posfrac(b[i]) > smallvalue THEN
    BEGIN  addconstraint(i);  stillinteger:=false  END
  END;
  integersolution := stillinteger
END; { integersolution }


PROCEDURE Simplexmod (p:phase);
VAR  n : integer;  unbounded, nonfeasible : boolean;

  PROCEDURE nextbasicvariable (VAR r,s:integer; x:row);
  VAR  i, j : integer;  min : real;
  BEGIN  min:=largevalue;  { Decide whether to employ Simplex }
    FOR j:=1 TO n DO        { Method or Dual Simplex Method.   }
      IF nonbasic[j] THEN IF x[j]<min THEN BEGIN min:=x[j]; s:=j END;
    solution := x[s] > -smallvalue;
    IF NOT solution THEN  { Find pivot in column s using Simplex Method }
    BEGIN  unbounded:=true;  i:=1;   { Check that at least one value }
      WHILE unbounded AND (i<=m) DO  { in column s is positive.      }
      BEGIN  unbounded := a[i,s] < smallvalue;  i:=i+1  END;
      IF NOT unbounded THEN     { Variable s enters the basis, now }
      BEGIN  min:=largevalue;  { Find the variable, basic[r], to }
        FOR i:=1 TO m DO        { leave the basis.                }
          IF a[i,s] > smallvalue THEN
            IF b[i]/a[i,s] < min THEN BEGIN min:=b[i]/a[i,s]; r:=i END;
      END
    END
    ELSE  { Find pivot using Dual Simplex Method }
    BEGIN  min:=largevalue;  { Find variable, basic[r], }
      FOR i:=1 TO m DO         { to leave basis.          }
        IF b[i]<min THEN BEGIN min:=b[i]; r:=i END;
      solution := b[r] > -smallvalue;
      IF NOT solution THEN
      BEGIN  min:=-largevalue;  { Find the variable, s, }
        FOR j:=1 TO n DO         { to enter the basis.   }
          IF nonbasic[j] THEN
            IF a[r,j] < -smallvalue THEN
              IF x[j]/a[r,j] > min THEN
```

```
                    BEGIN  min := x[j]/a[r,j];  s:=j  END;
          nonfeasible := min < -largevalue + smallvalue
      END
    END;
    IF NOT (solution OR unbounded OR nonfeasible) THEN
    BEGIN nonbasic[basic[r]]:=true; nonbasic[s]:=false; basic[r]:=s;
      writeln; writeln('     PIVOT IS AT ROW ', r:fwi, ' COL ', s:fwi)
    END
  END; { nextbasicvariable }


  PROCEDURE transformtableau (r,s:integer; VAR x:row; VAR x0:real);
  { See FullSimplex }

BEGIN  { Simplexmod }
  solution:=false; unbounded:=false; nonfeasible:=false;
  IF p=PhaseI THEN n:=n1 ELSE n:=n2;  { Determine current tableau size }
  REPEAT
    IF printon THEN outputtableau(p);
    CASE p OF
      PhaseI  : nextbasicvariable(r,s,d);
      PhaseII : nextbasicvariable(r,s,c)
    END;
    IF NOT (solution OR unbounded OR nonfeasible) THEN
      CASE p OF
        PhaseI  : transformtableau(r,s,d,w0);
        PhaseII : transformtableau(r,s,c,z0)
      END
  UNTIL solution OR unbounded OR nonfeasible ;
  IF unbounded THEN writeln('     UNBOUNDED')
  ELSE IF nonfeasible THEN writeln('     NO FEASIBLE SOLUTION')
END; { Simplexmod }


BEGIN  { Main Program }
  writeln; writeln('     GOMORY''S METHOD'); writeln;
  m:=mstart; inputdata; initialise; completetableau;
  IF GCplusEC=0 THEN writeln('     THERE IS NO PHASE I')
  ELSE { Perform Phase I }
  BEGIN  writeln('     PHASE I');
    Simplexmod(PhaseI); writeln;
    IF (abs(w0)>smallvalue) OR (NOT solution) THEN
    BEGIN  OK:=false; writeln('     PHASE I NOT COMPLETED');
         writeln('     SUM OF ARTIFICIALS ', w0:fwt,dpt)
    END
    ELSE
    BEGIN  writeln; writeln('     PHASE I SUCCESSFUL'); writeln;
      writeln('     REDUCED TABLEAU FOR PHASE II')
    END
  END;
  IF OK THEN  { Perform Phase II }
  BEGIN  Simplexmod(PhaseII); writeln;
    IF NOT solution THEN writeln('     PHASE II NOT COMPLETED')
    ELSE
```

```
      BEGIN
        WHILE NOT integersolution OR NOT solution DO Simplexmod(PhaseII);
        { Output final details }
        writeln; writeln('    FINAL SOLUTION'); writeln;
        writeln('    MINIMUM OF Z = ', -z0:fwt:dpt); writeln;
        writeln('    CONSTRAINT    BASIS    VALUE    STATE        SLACK');
        FOR i:=1 TO m DO slack[basic[i]] := b[i];
        FOR i:=1 TO m DO { For each constraint }
        BEGIN  write(i:10, basic[i]:10, ' ':12-fwt, b[i]:fwt:dpt, ' ':5);
          IF (i<=GC) OR (i>GCplusEC) THEN
            IF nonbasic[nvar+i] THEN writeln('BINDING', 0.0:10:dpt)
            ELSE  writeln('SLACK', ' ':12-fwt, slack[nvar+i]:fwt:dpt)
          ELSE  writeln('EQUATION    NONE')
        END
      END
    END
END. { Gomory }


GOMORY'S METHOD

THERE IS NO PHASE I

ITERATION   0
BASE VAR.      VALUE      X1      X2      X3      X4
   X3           5.00    1.00    1.00    1.00    0.00
   X4          10.00    1.00    5.00    0.00    1.00
   -Z           0.00   -2.00   -9.00    0.00    0.00

PIVOT IS AT ROW 2 COL 2

ITERATION   1
BASE VAR.      VALUE      X1      X2      X3      X4
   X3           3.00    0.80    0.00    1.00   -0.20
   X2           2.00    0.20    1.00    0.00    0.20
   -Z          18.00   -0.20    0.00    0.00    1.80

PIVOT IS AT ROW 1 COL 1

ITERATION   2
BASE VAR.      VALUE      X1      X2      X3      X4
   X1           3.75    1.00    0.00    1.25   -0.25
   X2           1.25    0.00    1.00   -0.25    0.25
   -Z          18.75    0.00    0.00    0.25    1.75


ITERATION   3
BASE VAR.      VALUE      X1      X2      X3      X4      X5
   X1           3.75    1.00    0.00    1.25   -0.25    0.00
   X2           1.25    0.00    1.00   -0.25    0.25    0.00
   X5          -0.75    0.00    0.00   -0.25   -0.75    1.00
   -Z          18.75    0.00    0.00    0.25    1.75    0.00

PIVOT IS AT ROW 3 COL 3
```

ITERATION 4

BASE VAR.	VALUE	X1	X2	X3	X4	X5
X1	0.00	1.00	0.00	0.00	-4.00	5.00
X2	2.00	0.00	1.00	0.00	1.00	-1.00
X3	3.00	0.00	0.00	1.00	3.00	-4.00
-Z	18.00	0.00	0.00	0.00	1.00	1.00

FINAL SOLUTION

MINIMUM OF Z = -18.00

CONSTRAINT	BASIS	VALUE	STATE	SLACK
1	1	0.00	SLACK	3.00
2	2	2.00	BINDING	0.00
3	3	3.00	BINDING	0.00

Example 1

Find non-negative integers x_1, x_2 which satisfy

$$3x_1 - 4x_2 \leqslant 12$$
$$x_1 + 2x_2 \leqslant 8$$

which minimise $-5x_1 - 2x_2$.

Full print out for this problem which has two 'less than or equal to' constraints in two variables is obtained by submitting the data file containing

$$1\ 0\ 0\ 2\quad 3\ -4\ 12\quad 1\ 2\ 8\quad -5\ -2.$$

The solution is given by $x_1 = 5$, $x_2 = 1$ as can be seen from the following output listing.

GOMORY'S METHOD

THERE IS NO PHASE I

ITERATION 0

BASE VAR.	VALUE	X1	X2	X3	X4
X3	12.00	3.00	-4.00	1.00	0.00
X4	8.00	1.00	2.00	0.00	1.00
-Z	0.00	-5.00	-2.00	0.00	0.00

PIVOT IS AT ROW 1 COL 1

ITERATION 1

BASE VAR.	VALUE	X1	X2	X3	X4
X1	4.00	1.00	-1.33	0.33	0.00
X4	4.00	0.00	3.33	-0.33	1.00
-Z	20.00	0.00	-8.67	1.67	0.00

PIVOT IS AT ROW 2 COL 2

ITERATION 2

BASE VAR.	VALUE	X1	X2	X3	X4
X1	5.60	1.00	0.00	0.20	0.40
X2	1.20	0.00	1.00	-0.10	0.30
-Z	30.40	0.00	0.00	0.80	2.60

ITERATION 3

BASE VAR.	VALUE	X1	X2	X3	X4	X5
X1	5.60	1.00	0.00	0.20	0.40	0.00
X2	1.20	0.00	1.00	-0.10	0.30	0.00
X5	-0.60	0.00	0.00	-0.20	-0.40	1.00
-Z	30.40	0.00	0.00	0.80	2.60	0.00

PIVOT IS AT ROW 3 COL 3

ITERATION 4

BASE VAR.	VALUE	X1	X2	X3	X4	X5
X1	5.00	1.00	0.00	0.00	0.00	1.00
X2	1.50	0.00	1.00	0.00	0.50	-0.50
X3	3.00	0.00	0.00	1.00	2.00	-5.00
-Z	28.00	0.00	0.00	0.00	1.00	4.00

ITERATION 5

BASE VAR.	VALUE	X1	X2	X3	X4	X5	X6
X1	5.00	1.00	0.00	0.00	0.00	1.00	0.00
X2	1.50	0.00	1.00	0.00	0.50	-0.50	0.00
X3	3.00	0.00	0.00	1.00	2.00	-5.00	0.00
X6	-0.50	0.00	0.00	0.00	-0.50	-0.50	1.00
-Z	28.00	0.00	0.00	0.00	1.00	4.00	0.00

PIVOT IS AT ROW 4 COL 4

ITERATION 6

BASE VAR.	VALUE	X1	X2	X3	X4	X5	X6
X1	5.00	1.00	0.00	0.00	0.00	1.00	0.00
X2	1.00	0.00	1.00	0.00	0.00	-1.00	1.00
X3	1.00	0.00	0.00	1.00	0.00	-7.00	4.00
X4	1.00	0.00	0.00	0.00	1.00	1.00	-2.00
-Z	27.00	0.00	0.00	0.00	0.00	3.00	2.00

FINAL SOLUTION

MINIMUM OF Z = -27.00

CONSTRAINT	BASIS	VALUE	STATE	SLACK
1	1	5.00	SLACK	1.00
2	2	1.00	SLACK	1.00
3	3	1.00	BINDING	0.00
4	4	1.00	BINDING	0.00

8.4   Dantzig's Method

Dantzig has also devised a cutting plane method for the solution of all integer linear programs. As with Gomory's method we first solve the problem without regard to the integer condition. If the problem has n variables $x_1, \ldots, x_n$ and the optimal basic feasible solution (ignoring the integer requirement) has basic variables $x_1, x_2, \ldots, x_m$, some of which are not integers, then the condition

$$x_{m+1} + x_{m+2} + \cdots + x_n \geqslant 1 \qquad (8.13)$$

is satisfied by all solutions with integer values but not by this basic solution.

It is clear that the basic solution does not satisfy (8.13) since $x_j = 0$ for $j = m + 1, \ldots, n$ and no *other* feasible solution exists with these particular non-basic variables all zero since they determine the values of the basic variables $x_1, x_2, \ldots, x_m$ uniquely. It follows then that for an all integer solution

$$x_{m+1} + x_{m+2} + \cdots + x_n \text{ must be positive.}$$

Since the x_j are to be integers this means that the sum must be at least one. Thus the additional constraint (and variable) takes the form

$$-x_{m+1} - x_{m+2} - \cdots - x_n + x_{n+1} = -1 \qquad (8.14)$$

where x_{n+1} is a non-negative integer.

As with Gomory's method we can proceed with the solution using the Dual Simplex Method. As an example to illustrate the method we consider again Example 1 of section 8.2:
find non-negative integers x_1, x_2 which satisfy

$$x_1 + x_2 \leqslant 5$$
$$x_1 + 5x_2 \leqslant 10$$

and minimise

$$-2x_1 - 9x_2 = z.$$

The computations as given by a computer program (see question 5 of Exercises 8) follow. The form of the added constraint is particularly simple and in contrast to Gomory's method does not involve fractions. However, it will also be observed that more cutting planes are needed before the optimal integer solution with $x_1 = 0$ and $x_2 = 2$ is obtained. Thus the cutting plane procedure of Gomory's method appears to be more efficient than the method of this section, and indeed is to be preferred.

```
DANTZIG'S METHOD

THERE IS NO PHASE I

ITERATION   0
BASE VAR.      VALUE      X1      X2      X3      X4
   X3           5.00     1.00    1.00    1.00    0.00
   X4          10.00     1.00    5.00    0.00    1.00
   -Z           0.00    -2.00   -9.00    0.00    0.00
```

PIVOT IS AT ROW 2 COL 2

ITERATION 1

BASE VAR.	VALUE	X1	X2	X3	X4
X3	3.00	0.80	0.00	1.00	-0.20
X2	2.00	0.20	1.00	0.00	0.20
-Z	18.00	-0.20	0.00	0.00	1.80

PIVOT IS AT ROW 1 COL 1

ITERATION 2

BASE VAR.	VALUE	X1	X2	X3	X4
X1	3.75	1.00	0.00	1.25	-0.25
X2	1.25	0.00	1.00	-0.25	0.25
-Z	18.75	0.00	0.00	0.25	1.75

ITERATION 3

BASE VAR.	VALUE	X1	X2	X3	X4	X5
X1	3.75	1.00	0.00	1.25	-0.25	0.00
X2	1.25	0.00	1.00	-0.25	0.25	0.00
X5	-1.00	0.00	0.00	-1.00	-1.00	1.00
-Z	18.75	0.00	0.00	0.25	1.75	0.00

PIVOT IS AT ROW 3 COL 3

ITERATION 4

BASE VAR.	VALUE	X1	X2	X3	X4	X5
X1	2.50	1.00	0.00	0.00	-1.50	1.25
X2	1.50	0.00	1.00	0.00	0.50	-0.25
X3	1.00	0.00	0.00	1.00	1.00	-1.00
-Z	18.50	0.00	0.00	0.00	1.50	0.25

ITERATION 5

BASE VAR.	VALUE	X1	X2	X3	X4	X5	X6
X1	2.50	1.00	0.00	0.00	-1.50	1.25	0.00
X2	1.50	0.00	1.00	0.00	0.50	-0.25	0.00
X3	1.00	0.00	0.00	1.00	1.00	-1.00	0.00
X6	-1.00	0.00	0.00	0.00	-1.00	-1.00	1.00
-Z	18.50	0.00	0.00	0.00	1.50	0.25	0.00

PIVOT IS AT ROW 4 COL 5

ITERATION 6

BASE VAR.	VALUE	X1	X2	X3	X4	X5	X6
X1	1.25	1.00	0.00	0.00	-2.75	0.00	1.25
X2	1.75	0.00	1.00	0.00	0.75	0.00	-0.25
X3	2.00	0.00	0.00	1.00	2.00	0.00	-1.00
X5	1.00	0.00	0.00	0.00	1.00	1.00	-1.00
-Z	18.25	0.00	0.00	0.00	1.25	0.00	0.25

```
ITERATION  7
BASE VAR.      VALUE     X1      X2      X3      X4      X5      X6      X7
   X1          1.25    1.00    0.00    0.00   -2.75    0.00    1.25    0.00
   X2          1.75    0.00    1.00    0.00    0.75    0.00   -0.25    0.00
   X3          2.00    0.00    0.00    1.00    2.00    0.00   -1.00    0.00
   X5          1.00    0.00    0.00    0.00    1.00    1.00   -1.00    0.00
   X7         -1.00    0.00    0.00    0.00   -1.00    0.00   -1.00    1.00
   -Z         18.25    0.00    0.00    0.00    1.25    0.00    0.25    0.00

PIVOT IS AT ROW 5 COL 6

ITERATION  8
BASE VAR.      VALUE     X1      X2      X3      X4      X5      X6      X7
   X1          0.00    1.00    0.00    0.00   -4.00    0.00    0.00    1.25
   X2          2.00    0.00    1.00    0.00    1.00    0.00    0.00   -0.25
   X3          3.00    0.00    0.00    1.00    3.00    0.00    0.00   -1.00
   X5          2.00    0.00    0.00    0.00    2.00    1.00    0.00   -1.00
   X6          1.00    0.00    0.00    0.00    1.00    0.00    1.00   -1.00
   -Z         18.00    0.00    0.00    0.00    1.00    0.00    0.00    0.25
```

FINAL SOLUTION

MINIMUM OF Z = -18.00

```
CONSTRAINT   BASIS    VALUE    STATE      SLACK
    1          1      0.00    SLACK       3.00
    2          2      2.00    BINDING     0.00
    3          3      3.00    SLACK       2.00
    4          5      2.00    SLACK       1.00
    5          6      1.00    BINDING     0.00
```

Exercises 8

1　Find non-negative integers x_1, x_2 which satisfy

$$7x_1 + 5x_2 \leqslant 35$$

$$x_1 + 2x_2 \leqslant 8$$

and minimise　　　　　$-6x_1 - 5x_2 = w.$

2　Use Gomory's method to find the minimum value of $z = x_1 - 11x_2$ where x_1 and x_2 are non-negative integers which satisfy

$$-x_1 + 10x_2 \leqslant 40$$

$$x_1 + \quad x_2 \leqslant 20.$$

3　The Travelling Salesman problem. A salesman has to visit n towns; the distance between town i and town j is d_{ij}. The salesman has to start at a given town and visit all the others just once and return to his starting point, in such a way that the total distance travelled is a minimum. (There are of course $(n - 1)!$ possible circular tours

so a solution by enumeration may seem possible, although hardly practical for large values of n.)
Formulate the problem as an all integer linear programming problem.

4 Find non-negative integers x_1, x_2, x_3 which satisfy

$$70x_1 + 60x_2 + 35x_3 \leqslant 215$$

and maximise $6x_1 + 5x_2 + 4x_3 = z.$

5 Modify the program given so as to implement Dantzig's method. (Note it is only necessary to remove the FUNCTION *posfrac* and change parts of the PROCEDURE *addconstraint*.) Use your program to solve questions 1, 2, 4. Compare the amount of computation with that used with Gomory's method.

6 Find non-negative x_1, x_2,..., x_5 such that

$$x_1 + \qquad 4x_3 + 2x_4 + x_5 \leqslant 41$$
$$4x_1 + 3x_2 + \quad x_3 - 4x_4 - x_5 \leqslant 47$$

which maximise

$$2x_1 + 3x_2 + 4x_3 + 2x_4 + x_5 = z$$

if the x_i are also constrained to be integers.

7 Find non-negative integers x_1, x_2, x_3 which minimise $4x_1 + 5x_2 + 2x_3 = w$ if

$$x_1 + 4x_2 \qquad \geqslant 11$$
$$3x_1 + 3x_2 + 2x_3 \geqslant 13$$
$$3x_1 + 2x_2 \qquad \geqslant 10.$$

8 It may have been noticed that degenerate solutions frequently arise when using Gomory's method. Write a program based on the Revised Simplex Method which will avoid the dangers of cycling and which will implement Gomory's method. [Note that the extra constraint will have to be incorporated into the matrix of original coefficients. Then the inverse of the basis and the simplex multipliers will have to be updated before we use the Dual Simplex Method to choose the variables to leave and enter the basis.] Try your program on the questions of these exercises.

9 A parcel is to be made up of items of two types; A and B. One item of type A weighs 4 kg., occupies 3 litres and has value $20. One item of type B weighs 7 kg., occupies 2 litres and has value $16. If the parcel must not exceed 28 kg. in weight nor 12 litres in volume and is to have maximum value, how many items of A and B should it contain?

10 One article of type A needs 4 kg. of raw material and 2 hours of machine time for its production. One article of type B needs 5 kg. of raw material and 1 hour of machine time for its production. Each half day session 20 kg. of raw material and 6 hours of machine time are available. The profit on one article of type A and B is $120 and $100 respectively. If each article started in a half day session has to be completed in that session find the most profitable production schedule.

Suggestions for Further Reading

Readers may be interested to consult some of the original work that has been discussed in this book. Also mentioned are some other textbooks which discuss methods and applications of Linear Programming.

Beale, E.M.L. Cycling in the Dual Simplex Algorithm, *Nav. Res. Logistics Quart.*, **2**, 269–276, 1955.

Dantzig, G.B. 'Maximisation of a linear function of variables subject to linear inequalities', in *Activity Analysis of Production and Allocation* (Edited by T. C. Koopmans), Wiley, New York, 1951.

Dantzig, G.B. *Linear Programming and Extensions*, Princeton University Press, New Jersey, 1963.

Dantzig, G.B. 'The Simplex Method', *Rand Corp. Rept.*, P-891, 1956.

Dantzig, G. B. and Orchard-Hays, W. 'Alternate Algorithm for the Revised Simplex Method Using Product Form for the Inverse', *Rand Corp. Rept.*, RM-1268, 1953.

Dantzig, G.B. 'Note on Solving Linear Programs in Integers', *Nav. Res. Logistics Quart.*, **6**, 75–76, 1959.

Ford, L.R., Jr. and Fulkerson, D.R. Solving the Transportation Problem, *Man. Sc.*, **3**, 24–32, 1956.

Gale, D. *The Theory of Linear Economic Models,* McGraw-Hill, New York, 1960.

Garvin, W. W. *Introduction to Linear Programming,* McGraw-Hill, New York, 1960.

Gass, S.I. *Linear Programming: Methods and Applications,* McGraw-Hill, 3rd Ed., New York, 1969.

Glover, F., Karney, D., Klingman, D. and Napier, A. A Computation Study on Start Procedures, Basis Change Criteria and Solution Algorithms for Transportation Problems, *Man. Sc.,* **20**, 793–813, 1974.

Gomory, R.E. 'Outline of an Algorithm for Integer Solutions to Linear Programs', *Bull. American Math. Soc.,* **64**, No. 5, 275–278, 1958.

Gomory, R. E. 'An Algorithm for Integer Solutions to Linear Programs,' *Recent Advances in Mathematical Programming,* (Edited by R. Graves and P. Wolfe), McGraw Hill, 269–302, 1963.

Hadley, G. *Linear Programming,* Addison Wesley, Reading, Mass., 1962.

Hitchcock, F. L. The Distribution of a Product from Several Sources to Numerous Localities, *J. Math. Phys.,* **20**, 224–230, 1941.

Hoffman, A. J. Cycling in the Simplex Algorithm, *Nat. Bur. Standards Rept.,* 2974, 1953.

Kuhn, H.W. The Hungarian Method for the Assignment Problem, *Nav. Res. Logistics Quart.,* **2**, 83–97, 1955.

Mack, C. The Bradford Method for the Assignment Problem, *The New J. of Stats. and Op. Res.,* **1**, Part 1, 17–29, 1969.

Orchard-Hays, W. *Advanced Linear Programming Computing Techniques*, McGraw-Hill, New York, 1968.

Wagner, H.M. A Comparison of the Original and Revised Simplex Methods, *Op. Res.,* **5**, 361–369, 1957.

Walsh, G. R. *An Introduction to Linear Programming,* Holt, Rinehart and Winston, 1971.

Solutions

Exercises 1

1 x_1 A items, x_2 B items; $x_1, x_2 \geqslant 0$.

Maximise $z = 5x_1 + 3x_2$ subject to $0.5x_1 + 0.25x_2 \leqslant 40$

$$0.4x_1 + 0.3x_2 \leqslant 36$$
$$0.2x_1 + 0.4x_2 \leqslant 36.$$

Optimum $x_1 = 60$, $x_2 = 40$, $z = \$420$ per week.

2   Maximum 10 when $x_1 = 2$, $x_2 = 4$.

4   Minimum -10 when $x_1 = 4$, $x_2 = 2$.

5   (a) Minimum -13, $x_1 = 3\frac{1}{2}$, $x_2 = 2\frac{1}{2}$.

    (b) Minimum -9, $x_1 = \dfrac{25 - 13\theta}{10}$, $x_2 = \dfrac{15 + 39\theta}{10}$; $0 \leqslant \theta \leqslant 1$.

    (c) Unbounded solution.

6   Contradictory constraints.

7   Maximum $\frac{12}{5}$, $x_1 = \frac{2}{5}$, $x_2 = \frac{1}{5}$, $x_3 = 0$.

8 $x_1 = \frac{1}{12}$, $x_2 = \frac{4}{12}$, $x_3 = \frac{7}{12}$, $C = \$38\frac{3}{4}$ per tonne.

9 x_1, x_2, x_3, proportions of A, B, C; $x_i \geqslant 0$.
Minimise $70x_1 + 50x_2 + 10x_3$ subject to

$$90x_1 + 65x_2 + 45x_3 \geqslant 60$$
$$30x_1 + 85x_2 + 70x_3 \geqslant 60$$
$$x_1 + x_2 + x_3 = 1.$$

Optimum $x_1 = \frac{17}{59}$, $x_2 = \frac{6}{59}$, $x_3 = \frac{36}{59}$.

10 $0 \leqslant x_1 \leqslant 30, 0 \leqslant x_2 \leqslant 20, 3x_1 + 3x_2 \leqslant 100$
Maximise $12(26x_1 + 45x_2)/5$
Optimum when $x_1 = \frac{40}{3}$, $x_2 = 20$ (tonnes).

12   A gets $(1, 4, 0)$ buses from G_1, G_2, G_3
      B gets $(2, 0, 5)$ buses from G_1, G_2, G_3
Total distance travelled is 25 miles.

13 $x_A = 450$, $x_B = 100$.

14   367 Caprice, 683 Fiasco.

15   Let x tonnes be transported from Leeds to Manchester, y tonnes from Leeds to Birmingham. Then the rest of the schedule will follow.
Optimum $x = 400$, $y = 400$.
In second case $x = 400$, $y = 100$.

Exercises 2

1   2.4: $x_1 = 0.4$, $x_2 = 0.2$, $x_3 = 0$.

2   0.8: $x_1 = 0$, $x_2 = 0.2$, $x_3 = 0.3$.

3 $x_A = 13.125$, $x_B = 2.625$, Profit = \$55.125.

4   40 000 $\frac{1}{2}$ litre bottles on A, 20 000 litre bottles on A, 10 000 litre bottles on B.

5 $z = 150$, $x_1 = 1$, $x_2 = 4$.

6 $w = 150$, $y_1 = 0$, $y_2 = \frac{25}{9}$, $y_3 = \frac{125}{9}$.

7 $x_i =$ No. of radiators of each type; $x_A = 400$, $x_B = 0$, $x_C = 150$, $x_D = 0$; Profit = \$3875.

8 $x_A = 2000$, $x_B = 7000$, Profit = \$10 350.

9 $x_1 = \frac{50}{7}$, $x_2 = 0$, $x_3 = \frac{55}{7}$, $z = -\frac{695}{7}$.

10   \$200 000 on television, \$100 000 on radio, \$100 000 on newspapers, \$100 000 on posters.

11 $z = -2$, $x_1 = 2$, $x_2 = 0$.

12   Min. cost 150: $\frac{5}{6}$ kg dried fish, 5 kg fruit, $3\frac{1}{3}$ litres milk sub.

13   Use 9000 gallons of W in A, 81 000 gallons of Y in A
         15 000 gallons of W in B, 85 000 gallons of Y in B
         24 000 gallons of X in C, 60 000 gallons of Y in C, 36 000 gallons of Z
             in C.

14   26 weavers; 4 start on 1st shift, 10 on 2nd, 8 on 4th, 4 on 5th, or 4 start on 1st, 4 on 2nd, 6 on 3rd, 1 on 4th, 11 on 5th.

15   No feasible solution.

Exercises 3

1   5500 tonnes of A, 4500 tonnes of B; \$100.

2 $x_1 = 160/11$, $x_2 = 60/11$, $x_3 = x_4 = 0$; $z = -460$.
         $x_1 = 10$, $x_2 = 5$, $x_3 = x_4 = 0$; $z = -455$.

4 $x_1 = 4$, $x_2 = 2$, $x_3 = 0$; $P = 24$. A, 0.6; B, 1.2.

5 $x_A = 18/5$, $x_B = 3/5$, $x_C = 0$; $P = 19\frac{4}{5}$. I, $\frac{7}{5}$; II, $\frac{1}{5}$.

6 Sawmill, 0; Assembly shop, \$1; Finishing shop, \$7
$x_1 = 150$, $x_2 = 50$, $x_3 = 10$.

7 2700 home, 2150 overseas. Yes.

8 (i) $\frac{3}{5}$, $\frac{1}{5}$ (ii) $\begin{pmatrix} -\frac{3}{5} & \frac{4}{5} \\ \frac{2}{5} & -\frac{1}{5} \end{pmatrix}$ (iii) Yes; $x_1 = 4$, $x_4 = 24$ (iv) $66\frac{2}{3} \leqslant P \leqslant 100$

(v) $x_1 = 11\frac{1}{4}$, $x_2 = 6\frac{1}{4}$, $x_4 = 15$.

9 $x_1 = 1$, $x_2 = 4$; $z = 150$.

10 No feasible solution.

12 $x_1 = \frac{22}{7}$, $x_2 = \frac{20}{7}$; $z = -\frac{144}{7}$. $x_1 = 0$, $x_2 = 4$, $z = -20$.

14 $x_1 = x_2 = 0$, $x_3 = 2$, $z = 4$.

15 E is obtained by deleting the last row and column from F.

Exercises 4

1 $x_{13} = 8$, $x_{21} = 2$, $x_{22} = 9$, $x_{31} = 2$, $x_{33} = 1$, $x_{34} = 13$; $C = 103$.

2 $x_{13} = 9$, $x_{22} = 4$, $x_{26} = 10$, $x_{31} = 6$, $x_{33} = 1$, $x_{34} = 2$, $x_{35} = 7$, $x_{44} = 11$, $x_{46} = 0$;
$C = 144$.

3 x_{ij} is the amount (1000 tonnes) sent to C_i from M_j. $x_{11} = 7$, $x_{41} = 36$, $x_{32} = 25$,
$x_{13} = 5$, $x_{23} = 15$.

4 x_{ij} is the number of posts in P_i filled from staff in S_j. Optimum is $x_{12} = 5$, $x_{14} = 1$,
$x_{15} = 2$, $x_{25} = 3$, $x_{31} = 2$, $x_{34} = 7$, $x_{43} = 4$, $x_{45} = 1$, with 23 satisfactory placings and
the two unsatisfactory placings ($x_{15} = 2$) in group s_5.

5 x_{ij} is the number of components (1000's) produced at F_i which are sent to C_j.
$x_{11} = 10$, $x_{14} = 30$, $x_{23} = 20$, $x_{31} = 5$, $x_{32} = 20$.

6 From F_1, 10 000 to C_1, 6000 to C_2; from F_2, 7000 to C_2, and 7000 to C_3, 2000
of the latter being produced by overtime working.

7 Airline I: 10 flights to Beirut, 10 to Dallas
Airline II;   10 flights to Sydney, 10 to Calcutta, 10 to Beirut
Airline III: 5 flights to Calcutta, 15 to Sao Paulo
Minimum cost = \$86 000.

8 I   sends   500 tonnes to C, 450 tonnes to E
II   sends   300 tonnes to D
III   sends 1000 tonnes to B, 350 tonnes to D
IV   sends   250 tonnes to A, 200 tonnes to C.

9   Plant I,   August, sends 420 to CI, 50 to CII and stores 30 which go to CI in September
Plant II, August, sends 300 to CII
Plant I,   September, sends 520 to CI (in addition to 30 stored from August)
Plant II, September, sends 480 to CII.

10   A   1500 to W, 2500 to Y
B   2500 to X, 200 to Y, 300 to Z
C   3000 to Z.

Exercises 5

1   (a) $x_{11} = 1, x_{23} = 1, x_{32} = 1$; Sum $= 9$
(b) $x_{11} = 1, x_{23} = 1, x_{32} = 1, x_{44} = 1$; Sum $= 37$.

2   (a) $x_{14} = 1, x_{21} = 1, x_{32} = 1, x_{43} = 1, x_{55} = 1$; Sum $= 100$
(b) $x_{15} = 1, x_{22} = 1, x_{31} = 1, x_{43} = 1, x_{54} = 1, x_{66} = 1$; Sum $= 158$.

3   S—D—d
T—B—b or T—E—e
U—C—c
V—A—a
W—E—e or W—B—b.

4   P1—A3, P2—A2, P3—A4, P4—A1.

5   Wednesday a.m. in A, Wednesday p.m. in D, Thursday a.m. in C, Thursday p.m. in B, Friday a.m. in E.

6   At C: 2—12, 6—11
At B: 1—9, 3—10, 4—7, 5—8
At A: 7—4, 11—5, 8—6, 9—1, 12—2, 10—3.

7   Task 1 for op. 5, task 2 for op. 4, task 3 for op. 1, task 4 for op 7, task 5 for op. 2, task 6 for op. 3, task 7 for op. 6. Time 111 hours.

8 $A \rightarrow I, B \rightarrow II, C \rightarrow III, D \rightarrow IV, E \rightarrow V$.

Exercises 6

1 $x_1 = 300, x_2 = 200, z = -1400$.

2 $x_1 = 12, x_2 = 8, z = 68$.

3 $x_1 = 1, x_2 = 4, z = 150$.

4 $y_1 = 0, y_2 = \frac{25}{9}, y_3 = \frac{125}{9}, w = 150$.

5 $x_1 = 1, x_2 = 0, x_3 = 1, x_4 = 0, z = -1.25$.

6   Component production only. New schedule $x_1 = 90, x_2 = 30, x_3 = 30$.

7 $x_1 = 50, x_2 = 50, x_3 = 50$; Profit = \$2500.
$x_1 = 100, x_2 = 150, x_3 = 0$; Profit = \$3000.
$x_1 = 75, x_2 = 100, x_3 = 25$; Profit = \$2775.

8 Grow 10×10^5 acres crop 1 on land I, 4×10^5 acres crop 2 on land I, 12×10^5 acres crop 3 on land II. All land and water resources are fully used but labour is underused by 10^5 men, over 14% of the work force available.
Each 10^5 acres of land I will yield 2 extra units of revenue.
Each 10^5 units of water will yield 2 extra units of revenue.
Yes. It increases revenue by $11\frac{1}{3}$ units (from 152 to $163\frac{1}{3}$). In the new solution grow $\frac{28}{3} \times 10^5$ acres of crop 1 on land I, $\frac{14}{3}$ acres of crop 2 on land I and $\frac{35}{3}$ acres of new crop on land II. All labour is now utilised but not all land II.

9

	Beaujolais	Nuits St Georges	St Emilion
French burgundy	54	46	
French claret	36	$7\frac{2}{3}$	$86\frac{1}{3}$
Spanish red	90	23	37

Results expressed in 1000's bottles.
Profit = £490 133.33.

Exercises 7

1 Find $y_1, y_2, y_3 \geqslant 0$ such that

$$3y_1 + y_2 + 2y_3 \leqslant 11$$
$$5y_1 + 9y_2 + 7y_4 \leqslant 44$$

which maximise $\qquad 18y_1 + 30y_2 + 27y_3$.

2 Find y_1, y_2, y_3 where $y_1, y_3 \geqslant 0$ and y_2 is unrestricted such that

$$2y_1 + y_2 + y_3 \leqslant 5$$
$$y_1 + y_2 + 2y_3 \leqslant 7$$
$$4y_1 + y_2 + 3y_3 = 13$$

which maximise $\qquad 22y_1 + 9y_2 - 18y_3$.

5 $x_1 = 1, x_2 = 2, x_3 = 3$; with $z_{\min} = 84$; $\pi_1 = -3$, $\pi_2 = -2$, $\pi_3 = -1$. For dual, $y_1, y_2, y_3 \geqslant 0$

$$2y_1 + y_2 + 3y_3 \leqslant 11$$
$$3y_1 + 2y_2 + y_3 \leqslant 14$$
$$y_1 + 5y_2 + 2y_3 \leqslant 15$$
$$11y_1 + 20y_2 + 11y_3 = w \text{ is to be maximised.}$$

$y_1 = 3, y_2 = 2, y_3 = 1$, with $w_{\max} = 84$; $\rho_1 = -1, \rho_2 = -2, \rho_3 = -3$.

6 Dual is find $y_1, y_2 \geqslant 0$

such that
$$y_1 \qquad \leqslant -1$$
$$-y_1 - y_2 \leqslant -1$$

which maximise
$$y_1 - 2y_2.$$

This has no feasible solution.

8 Primal, find $x_1, x_2 \leqslant 0$

such that
$$-5x_1 + 5x_2 \geqslant -4$$
$$x_1 - x_2 \geqslant 2$$

which minimise $\quad -7x_1 + 3x_2$

has no feasible solution.

Dual, find $y_1, y \geqslant 0$

such that
$$-5y_1 + y_2 \leqslant -7$$
$$5y_1 - y_2 \leqslant 3$$

which maximise $\quad -4y_1 + 2y_2$

also has no feasible solution.

9 $x_1 = 4$, $x_2 = 1$, $x_3 = 0$; $z = 32$; $\pi_1 = 0$, $\pi_2 = -2$, $\pi_3 = -1$.

Exercises 8

1. Min $w = -30$ when $x_1 = 5$, $x_2 = 0$.

2. Min $z = -45$, $x_1 = 10$, $x_2 = 5$.

4. Max $z = 24$, when $x_1 = 0$, $x_2 = 0$, $x_3 = 6$.

6. Max $z = 167$ when $x_1 = 0$, $x_2 = 42$, $x_3 = 0$, $x_4 = 20$, $x_5 = 1$.

7. Min $w = 22$ when $x_1 = 3$, $x_2 = 2$, $x_3 = 0$.

9. 4 items of A, 0 items of B.

10. Make 2 items of A, 2 items of B per half day session.

Index

Addition of extra constraints, 63
Artificial objective function, 32, 126
  simplex multipliers, 126
Artificial variable, 31, 126
Assignment problem, 105
  Mack's method, 106
  program, 111

Basic feasible solution, (b.f.s.), 10, 20
  generating one, 31
Basic solution, 10
  for transportation problem, 78
Basic variable, 10
Basis, 10
  inverse of, 54, 121
  triangular for transportation problem, 79, 80

Canonical form, 20
Changes in the b_i, 58
Changes in the c_j, 61
Complementary slackness, 157
Constraints, 2
Contour lines, 2
Convex hull, 11
Convex set, 11
Corner, 11
Cutting plane, 169, 176
Cycling, 46, 128
  perturbation method for avoiding, 128–130

Dantzig's method, for full Integer Linear Programming, 176
Degeneracy, 43, 128
  in the transportation problem, 84
Duality, 148
  and Kuhn-Tucker theory, 160
Dual problem, 148
  for transportation problem, 159

Dual Simplex Method, 63, 167
  program, 66
  rules, 65
Duality theorems, 151

Extreme point, 11

Feasible region, 2
Feasible solution, 2, 9
Flow chart, for Simplex Method, 27
  for transportation problem, 90
  for transportation problem, first b.f.s., 91
  for transportation problem, u's and v's, 92
Fundamental results, for L.P. 11 et seq.

Gomory's method, for all Integer Linear Programming, 166 et seq.
  program, 170
Graphical solution, of 2-D problems, 4

Inclusion of extra variables, 62
Integers, Linear Programming in, 164 et seq.
Inverse of basis, 54, 121
Iterative process of the Simplex Method, 22

Kuhn-Tucker theory, relation of Duality, 160

Linear Programming in integers, 164 et seq.
Linear Programming problem, standard form, 8
Line segment, 11

Mack's method, for assignment problem, 106

Non-basic variable, (n.b.v.), 10
Non-negativity condition, 2

Objective function, 1

Pivot, column, 23
　element, 23
　row, 23
Primal problem, 148
Program, for assignment problem, 111
　for Dual Simplex Method, 66
　for Gomory's method, 170
　for Revised Simplex Method, 132
　for Simplex Method, 37
　for Simplex Method, from a b.f.s., 26
　for transportation problem, 93

Revised Simplex Method, 120
　initiating the method, 126
　program, 132

Sensitivity analysis, 54 et seq.

Shortest route, through a network, 164
Simplex Method, 35
　given a b.f.s., 19 et seq.
　program, 37
　program, given a b.f.s., 26
Simplex Multipliers, 56, 121, 123
　for transportation problem, 81
Simplex tableau, 22
Slack variable, 8, 9
Standard form, for L.P. problem, 8
Stepping stones algorithm, 81

Tableau Simplex, 22
Transportation array, 78
Transportation problem, 76
　dual, 159
　program, 93

Unbalance in transportation problem, 84
Unbounded solution, 7

Vertex, 11